MSC-01372-1

 NATIONAL AERONAUTICS AND SPACE ADMINISTRATION

APOLLO OPERATIONS HANDBOOK EXTRAVEHICULAR MOBILITY UNIT

MARCH 1971

VOLUME I
SYSTEM DESCRIPTION
CSD-A-789-(1)
APOLLO 15-17

CREW SYSTEMS DIVISION
ORIGINAL ISSUE AUGUST 1968
REVISION V

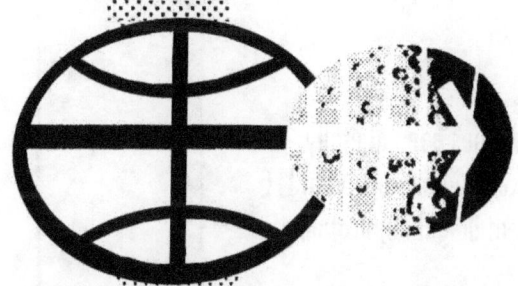

MANNED SPACECRAFT CENTER
HOUSTON, TEXAS

©2012 Periscope Film LLC
All Rights Reserved
ISBN # 978-1-937684-86-0
www.PeriscopeFilm.com

PROJECT DOCUMENT CHANGE/REVISION LOG FOR CSD ORIGINATED DOCUMENT NUMBER _____				PAGE i OF xii
CHG. NO. / DATE	AUTHORITY FOR CHANGE	PAGES AFFECTED	BRIEF DESCRIPTION OF CHANGE	END ITEM/ SERIAL NUMBER AFFECTED
Revision V 2/25/71	[signature] per W.W. Guy	All	Reorganized and rewritten to accomodate A7LB suit configuration and the -7 PLSS configuration	J Missions
Amend 1 7/7/71	[signature Gibson]	2-53, 2-54 2-57, 2-66 2-67, 2-78 2-79, 2-80 2-83, 2-85	To make technical changes and additions	Apollo 15-17
Amend 2 11/8/71	[signature Gibson]	2-3, 2-4 2-5, 2-6 2-7, 2-8 2-9, 2-10 2-11, 2-17 2-32, 2-35 2-38, 2-43 2-48, 2-49 2-52, 2-113, 2-116, 3-1, 3-3	To make technical changes and additions	Apollo 16-17

ALTERED PAGES MUST BE TYPED & DISTRIBUTED FOR INSERTION

MSC Form 892 (Rev Apr 69)

APOLLO OPERATIONS HANDBOOK

EXTRAVEHICULAR MOBILITY UNIT

VOLUME I — SYSTEM DESCRIPTION

CSD-A-789-(1)

Prepared by: _____ 8/2/68
James L. Gibson
Apollo Support Branch

Approved by: _____ 8/5/68
Charles C. Lutz, Chief
Apollo Support Branch

_____
Richard S. Johnston, Chief
Crew Systems Division

AUTHORIZED FOR DISTRIBUTION

_____
Maxime A. Faget
Director of Engineering and Development

NATIONAL AERONAUTICS AND SPACE ADMINISTRATION

MANNED SPACECRAFT CENTER

HOUSTON, TEXAS

July 1968

PREFACE

This document is the fifth revised issue of Volume I of the Apollo Operations Handbook. This revision incorporates applicable portions of revisions I, II, III, and IV, and reorganizes the presentation for the Apollo J missions.

CONTENTS

Section		Page
1.0	INTRODUCTION	1-1
1.1	PURPOSE	1-1
1.2	SCOPE	1-1
2.0	EXTRAVEHICULAR MOBILITY UNIT SUBSYSTEMS AND ACCESSORIES	2-1
2.1	GENERAL DESCRIPTION	2-1
2.2	FIELD OPTIONAL ITEMS	2-4
2.3	PRESSURE GARMENT ASSEMBLIES AND ACCESSORIES	2-7
2.3.1	EV A7LB Pressure Garment Assembly	2-14
2.3.2	CMP A7LB Pressure Garment Assembly	2-31
2.3.3	Interface Components	2-38
2.3.4	Controls and Displays	2-58
2.3.5	Pressure Garment Accessories	2-58
2.4	INFLIGHT COVERALL GARMENT	2-80
2.5	PORTABLE LIFE SUPPORT SYSTEM	2-82
2.5.1	Oxygen Ventilating Circuit	2-82
2.5.2	Primary Oxygen Subsystem	2-86
2.5.3	Liquid Transport Loop	2-89
2.5.4	Feedwater Loop	2-90
2.5.5	Electrical Power Subsystem	2-93
2.5.6	Extravehicular Communications System	2-95
2.5.7	Remote Control Unit	2-100

Section		Page
2.6	OXYGEN PURGE SYSTEM	2-105
2.7	BUDDY SECONDARY LIFE SUPPORT SYSTEM	2-109
2.8	PRESSURE CONTROL VALVE	2-113
2.9	PLSS FEEDWATER COLLECTION BAG	2-113
2.10	BIOMEDICAL INSTRUMENTATION SYSTEM	2-116
2.10.1	Electrocardiogram Signal Conditioner	2-116
2.10.2	Impedance Pneumogram Signal Conditioner	2-116
2.10.3	The dc-dc Power Converter	2-116
2.10.4	Electrodes	2-116
3.0	EXTRAVEHICULAR MOBILITY UNIT SYSTEMS	3-1
3.1	PRIMARY PRESSURIZATION AND VENTILATION	3-1
3.2	LIQUID COOLING SYSTEM	3-3

TABLES

Table		Page
2-I	EMU OPERATIONAL SPECIFICATIONS	2-3
2-II	FIELD OPTIONAL ITEMS	2-5
2-III	EV A7LB PRESSURE GARMENT ASSEMBLY AND ACCESSORIES INTERFACE CONFIGURATIONS	2-10
2-IV	CMP A7LB PRESSURE GARMENT ASSEMBLY AND ACCESSORIES INTERFACE CONFIGURATIONS	2-11
2-V	EV A7LB ITMG MATERIALS CROSS SECTION (LISTED FROM THE INSIDE OUT)	2-18
2-VI	MATERIALS CROSS SECTION FOR EV THERMAL GLOVE	2-26
2-VII	MATERIALS CROSS SECTION FOR LUNAR BOOT	2-31
2-VIII	CMP A7LB CLA MATERIALS CROSS SECTION (LISTED FROM THE INSIDE OUT)	2-36
2-IX	PERFORMANCE CHARACTERISTICS OF THE LIQUID COOLING GARMENT AND MULTIPLE WATER CONNECTOR	2-64
2-X	PLSS/EVCS CURRENT LIMITER RATINGS	2-96
2-XI	PLSS/EVCS COMMUNICATIONS TELEMETRY CHARACTERISTICS	2-101

FIGURES

Figure		Page
2-1	Lunar surface configuration of the extravehicular mobility unit	2-2
2-2	CMP A7LB pressure garment assembly and accessories interface configurations	2-8
2-3	EV A7LB pressure garment assembly and accessories interface configurations	2-9
2-4	EV A7LB integrated torso limb suit assembly	2-16
2-5	Pressure helmet assembly and helmet shield	2-21
2-6	Glove assemblies with wristlets	2-24
2-7	Detachable pocket assemblies	2-27
2-8	Biomedical harness and sensors	2-29
2-9	Lunar boots	2-30
2-10	Neck dam	2-32
2-11	CMP A7LB integrated torso limb suit	2-34
2-12	PLSS attachments	2-39
2-13	Lunar module tether attachments (A7LB EV)	2-40
2-14	Helmet attaching neck ring	2-41
2-15	Wrist disconnects	2-44
2-16	Gas connectors and diverter valve	2-45
2-17	Multiple water connector	2-48
2-18	Urine transfer connector	2-49
2-19	Medical injection patch	2-51
2-20	Zipper lock assemblies	2-52

Figure		Page
2-21	Pressure relief valve	2-54
2-22	Biomedical and suit electrical harness and biomedical belt	2-56
2-23	Fecal containment subsystem and urine collection and transfer assembly	2-59
2-24	Constant wear garment and electrical harness	2-61
2-25	Liquid cooling garment and adapter interconnect	2-63
2-26	Insuit drinking device	2-67
2-27	Communications carrier	2-69
2-28	Lunar extravehicular visor assembly	2-70
2-29	Dual-position purge valve	2-74
2-30	Inflight helmet stowage bag	2-76
2-31	LEVA helmet stowage bag	2-77
2-32	EMU maintenance kit	2-79
2-33	Inflight coverall garment	2-81
2-34	Portable life support system	2-83
2-35	Duration of -7 PLSS expendables	2-84
2-36	Schematic of PLSS	2-85
2-37	Oxygen ventilating circuit	2-87
2-38	Primary oxygen subsystem	2-88
2-39	Liquid transport loop	2-91
2-40	PLSS feedwater loop	2-92
2-41	Battery locking device	2-94

Figure		Page
2-42	Extravehicular communications system	
	(a) The EVC-1	2-97
	(b) The EVC-2	2-98
2-43	Remote control unit	
	(a) Pictorial view of main elements	2-102
	(b) Oxygen quantity indicator markings and accuracies	2-103
	(c) Dimensions	2-104
2-44	Oxygen purge system, -3 configuration	2-106
2-45	The OPS worn in the helmet-mounted mode	2-107
2-46	The OPS worn in the torso-mounted contingency mode	2-108
2-47	Oxygen purge system schematic	2-110
2-48	Buddy secondary life support system schematic	2-111
2-49	Buddy secondary life support system	2-112
2-50	BSLSS hose stowage	2-114
2-51	Pressure control system	2-115
2-52	Biomedical instrumentation system	2-117
3-1	EMU primary pressurization and ventilation system	3-2
3-2	EMU liquid cooling system	3-4

ACRONYMS

AM	amplitude modulation
BSLSS	buddy secondary life support system
CLA	cover layer assembly
CMP	command module pilot
CWG	constant wear garment
DV	diverter valve
ECG	electrocardiogram
ECS	environmental control system
EMU	extravehicular mobility unit
EV	extravehicular
EVA	extravehicular activity
EVC	extravehicular communicator
EVCS	extravehicular communications system
FCS	fecal containment subsystem
FM	frequency modulation
IHSB	inflight helmet stowage bag
IRIG	interrange instrument group
ITLSA	integrated torso limb suit assembly
ITMG	integrated thermal micrometeoroid garment
IV	intravehicular
LCG	liquid cooling garment
LEVA	lunar extravehicular visor assembly
LM	lunar module

MWC	multiple water connector
OPS	oxygen purge system
PCV	pressure control valve
PGA	pressure garment assembly
PHA	pressure helmet assembly
PLSS	portable life support system
RCU	remote control unit
TLSA	torso limb suit assembly
UCD	urine collection device
UCTA	urine collection and transfer assembly
UV	ultraviolet
ZPN	impedance pneumogram

1.0 INTRODUCTION

1.1 PURPOSE

This volume provides familiarization information essential to the operation of the extravehicular mobility unit (EMU), and describes the configuration combinations for the A7LB separable-components and the accessory contract end items. Configuration deviations may be made as dictated by specific crew/mission requirements. Operational procedures and malfunction detection procedures are found in Volume II of this handbook.

1.2 SCOPE

The descriptive information for the EMU subsystems and related components is given in section 2.0. A description of the EMU systems is provided in section 3.0.

2.0 EXTRAVEHICULAR MOBILITY UNIT SUBSYSTEMS AND ACCESSORIES

2.1 GENERAL DESCRIPTION

The EMU (fig. 2-1) is designed to protect the crewman in a low-pressure, micrometeoroid, and thermal environment and to provide comfort, mobility, dexterity, and a specified unobstructed range of vision during lunar-surface or free-space operations outside of the spacecraft. The EMU (table 2-I) provides the extravehicular (EV) crewman with a habitable environment for a 5-hour design mission without replenishment of expendables (based upon a 1200-Btu/hr metabolic rate with a 300-Btu/hr heat-leak rate).

There are two basic pressure garment assembly (PGA) configurations which support Apollo missions. One configuration is designated as the command module pilot (CMP) A7LB PGA which provides low-pressure and fire protection in the intravehicular (IV) mode and protection from the free-space environment during extravehicular activity (EVA) from the command module. The second configuration is designated as the EV A7LB PGA which provides low-pressure and fire protection in the IV mode and protection from the lunar surface environment during EVA. The EV A7LB PGA also provides free-space environment protection during open-hatch operations associated with command module (CM) EVA. Exterior connectors permit both configurations to interface with spacecraft systems for pressurization, ventilation, communications, cooling, and waste management. The EV configuration interfaces with the portable life support system (PLSS) for pressurization, ventilation, communications, and temperature control when used for EVA. The CMP A7LB PGA interfaces with the command service module (CSM) EVA umbilical assembly, the oxygen purge system (OPS), the purge valve, and the pressure control valve (PCV). Waste management systems are also self-contained in both configurations to permit operations while independent of the spacecraft waste management system.

2.2 FIELD OPTIONAL ITEMS

The items designated as crew/mission requirement deviations are shown in table 2-II. These items may be altered at the option of the individual crewman. Certain items are also adjustable as necessary to satisfy crewman comfort requirements. The deviations are determined as much as possible during the initial fit check; however, field modifications are accomplished when they are within the capability of the applicable support activity.

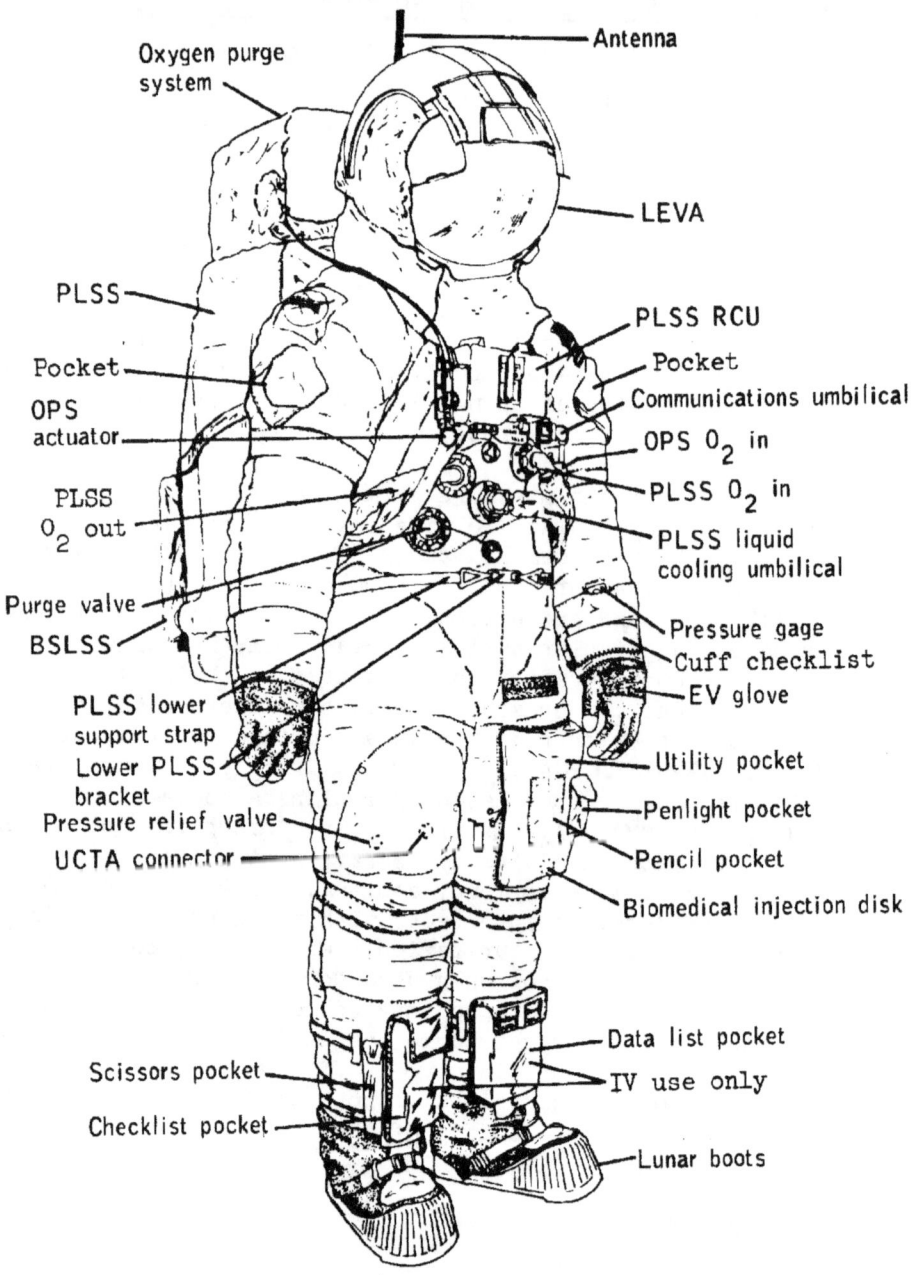

Figure 2-1.- Lunar surface configuration of the extravehicular mobility unit.

TABLE 2-I.- EMU OPERATIONAL SPECIFICATIONS

Item	Value
Pressure garment assembly	
Operational temperature limitations	-290° to +300° F
Leak rate at 3.7 psia (max.)	180.00 scc/min (0.0315 lb/hr)
Operating pressure	3.75 ± 0.25 psid
Structural pressure	6.00 psid
Proof pressure	8.00 psid
Burst pressure	10.00 psid
Pressure drop	
12 acfm, 3.5 psia, 50° F, and inlet diverter valve open (IV position)	4.70 in water
6 acfm, 3.9 psia, 77° F, and inlet diverter valve closed (EV position)	1.80 in water
Pressure gage range	2.5 to 6.0 psid
Pressure relief valve	
Cracking pressure	5.00 to 5.75 psid
Reseat pressure	4.6 psid min.
Suit pressure	5.85 psid max.
Leak rate closed	4.0 scc/min max.
Flow rate open	12.2 lb/hr min. of O_2 at 5.85 psia

Amendment 2
11/5/71

TABLE 2-I.- EMU OPERATIONAL SPECIFICATIONS - Concluded

Item	Value
Liquid cooling garment	
Operating pressure	4.20 to 23.0 psid
Structural pressure	31.50 ± 0.50 psid
Proof pressure	31.50 ± 0.50 psid
Burst pressure	47.50 psid
Pressure drop	
4.0 lb/min at 45° F inlet	3.35 psi including both halves of multiple water connector
Leak rate	
19.0 psid at 45° F	0.58 cc/hr
Multiple water connector	
Pressure drop	
4.0 lb/min at 45° F, both halves, both directions	1.45 psi
Portable life support system	
Oxygen quantity	145 to 1500 psia
Low oxygen flow	0.07 lb/hr
Low PGA pressure	3.10 to 3.40 psid
Low vent flow	4.0 acfm (min. at 15 mm Hg)
Carbon dioxide production	0.39 lb/hr
Low feedwater	1.2 to 1.7 psia

Amendment 2
11/5/71

TABLE 2-II.- FIELD OPTIONAL ITEMS

Item	Action
Leg mobility straps	Leg mobility straps may be removed.
Location of strap-on pockets	Strap-on pockets may be located as preferred by individual crewman.
Liner comfort pads	Comfort pads may be positioned as necessary to decrease pressure points.
Custom length of palm restraint straps	Palm restraint strap length may be varied to correspond with hand size.
Pocket preference for neck dam lanyard attaching strap	The neck dam lanyard strap may be stored to suit the individual crewman.
Orientation of gas connector locks	Gas connectors may be rotated to locate the locking tabs at 60° intervals to accommodate interface or operational requirements.
Custom length PGA urine drain hose	Hose length can be varied as necessary to accommodate fit.
Orientation or length of PGA liner electrical harness keeper tabs	Electrical harness keeper tabs may be lengthened or reorientated as necessary.
Wristlets	Wristlets may be donned as necessary to enhance crew comfort in wrist disconnect area.
Valsalva device	The valsalva device may be deleted from the pressure helmet at the discretion of the crewman.
Comfort gloves	The comfort gloves may be deleted.
Contingency sample pocket	The data list pocket includes a removable wall stiffener and is used as a contingency sample pocket during lunar surface activities.
Chin comfort pad	Comfort pads may be installed in the ITLSA liner for crewman comfort.

Amendment 2
11/5/71

TABLE 2-II.- FIELD OPTIONAL ITEMS - Concluded

Item	Action
Scissors pocket	The scissors pocket may be attached to the straps of the checklist pocket or the outer shell of the integrated thermal micrometeoroid garment (ITMG) adjacent to the utility pocket.
Limb adjustments	The arm and leg lengths may be adjusted to customize the lengths to the crewman.
Neck restraint guide	The neck restraint cable guide may be located in one of three positions to accommodate suit posture and crewman comfort.
Wrist disconnect comfort pads	Comfort pads may be installed within the wrist disconnect to preclude chafing and buffeting discomforts.
EVA checklist	A lunar surface EVA checklist may be attached to the EV glove gauntlet outer shell as a crew/mission requirement. The specific location, method of attachment, and orientation of the checklist on the glove gauntlet will be defined by the crewman to to satisfy his specific needs and mission objectives.
Vertical location of liquid cooling garment (LCG) manifold	The LCG manifold may be raised or lowered to provide maximum comfort.
Comfort pads for the LCG at shoulders and hips	Comfort pads may be installed on the LCG at the shoulders and/or hip areas as preferred by the crewman for his comfort.
LCG comfort modification	The LCG may be modified by adding or removing material to accommodate crewman size.
LCG turtleneck addition	A turtleneck collar may be donned with the LCG for additional comfort.

Amendment 2
11/5/71

2.3 PRESSURE GARMENT ASSEMBLIES AND ACCESSORIES

The Apollo pressure garment assemblies are anthropomorphic, protective structures worn by the crewmen during EV phases of an Apollo mission, and during IV modes of spacecraft operations. The CMP A7LB pressure garment configuration (fig. 2-2) is worn by the CMP and is normally used for IV and free space EV operations. The EV A7LB configurations (fig. 2-3) are worn by the crew commander and the lunar module (LM) pilot for IV and free space operations and lunar explorations.

The EV A7LB pressure garment and accessory systems interface with the portable life support systems to provide life support during lunar exploratory missions. The spacecraft environmental control EVA umbilical assembly and communications systems interface with the CMP A7LB pressure garment and accessories for free space EVA. Both configurations interface with the spacecraft crew systems and perform life support functions during depressurized and emergency modes of IV operations. The pressure garments permit normal body movements for the operation of spacecraft controls and equipment and have specially constructed devices required for space exploration. The garments are designed to operate at 0.18-psi (vent) to 3.75-psi (regulated) differential pressure at gas (oxygen) flow rates of 6 to 12 cubic feet per minute. The pressure garments are operational in temperatures of -290° to +300° F and in micrometeoroid flux densities normally expected within the lunar orbit perimeter. They can be worn for 115 hours during pressurized modes of emergency operation or 14 days of unpressurized operation except for normal removals for hygiene requirements. The pressurizable portion of the PGA includes an integrated torso limb suit assembly (ITLSA), detachable gloves, and a pressure helmet assembly (PHA). Entry into the EV A7LB torso limb suit is made through slide fastener (zipper) openings in the waist area. Entry into the CMP A7LB torso limb suit is gained through pressure-sealing and restraint-slide-fastener closures mounted vertically along the back and through the crotch. The helmet and gloves are then mechanically locked in place to complete the airtight envelope. Figure 2-2 and table 2-III identify the components that are interfaced for CMP A7LB EV and IV use, and figure 2-3 and table 2-IV identify the components interfaced to comprise EV A7LB suit configurations for normal EV and IV use.

Amendment 2
11/5/71

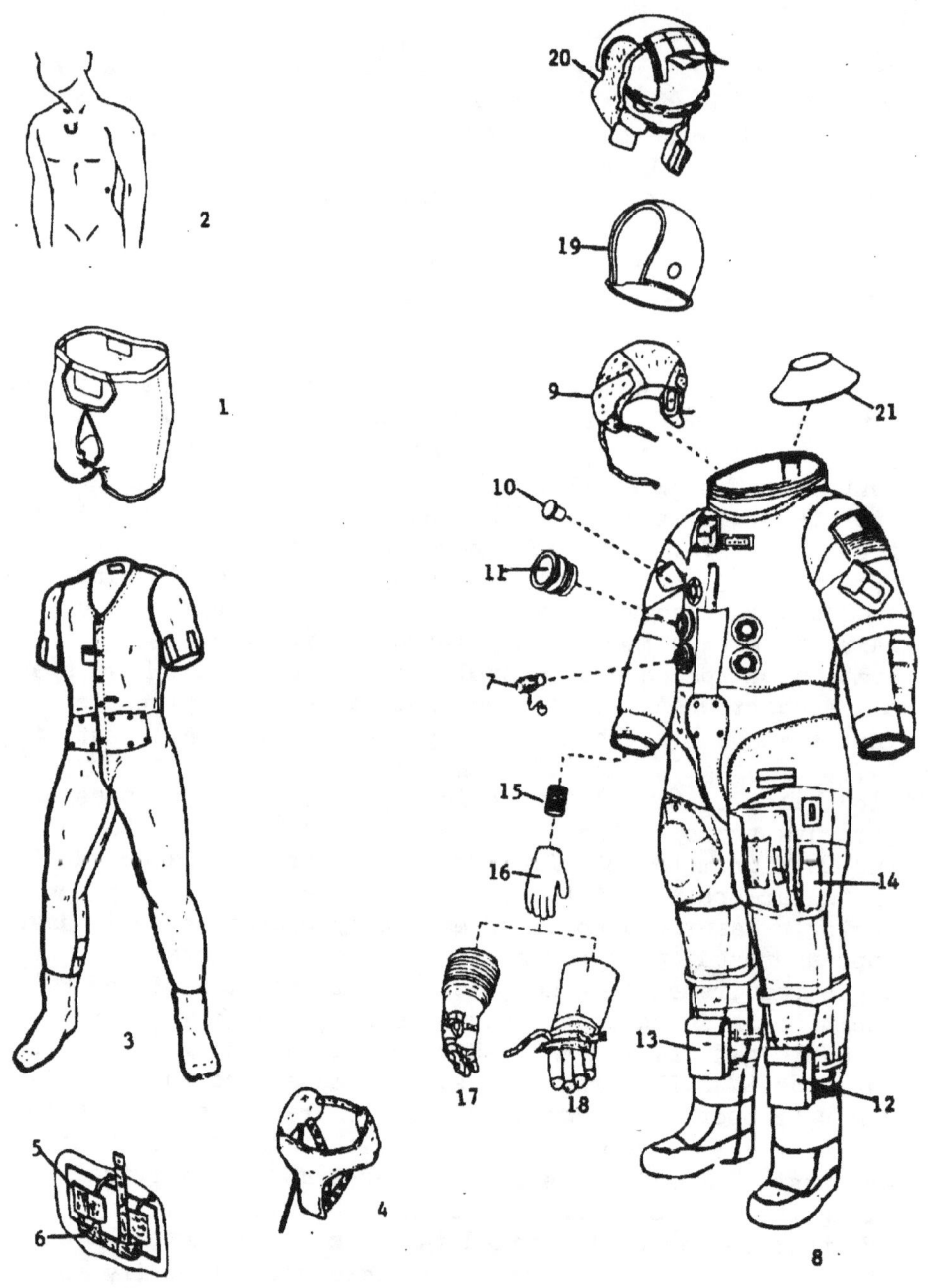

Figure 2-2.— CMP A7LB pressure garment assembly and accessories interface configurations.

TABLE 2-III.- CMP A7LB PRESSURE GARMENT ASSEMBLY AND ACCESSORIES
INTERFACE CONFIGURATIONS

Components	Use	
	EV	IV
1. Fecal containment subsystem	X	X
2. Biomedical sensors	X	X
3. Constant wear garment		X
4. Urine collection and transfer assembly	X	X
5. Biomedical belt	X	X
6. Biomedical harness	X	X
7. Purge valve	X	
8. EV integrated torso limb suit assembly	X	X
9. Communications carrier	X	X
10. Electrical connector cap		X
11. Gas connector caps		X
12. Data list pocket		X
13. Checklist pocket		X
14. Scissors pocket (attached to strips of checklist pocket or cover layer assembly shell outboard of and adjacent to the utility pocket)	X	X
15. Wristlets	X	X
16. Comfort gloves	X	X
17. IV pressure gloves		X
18. EV glove assemblies (used in place of IV pressure gloves for EV use)	X	
19. Pressure helmet assembly	X	
20. Lunar extravehicular visor assembly	X	
21. Neck dam (for water egress)		

Amendment 2
11/5/71

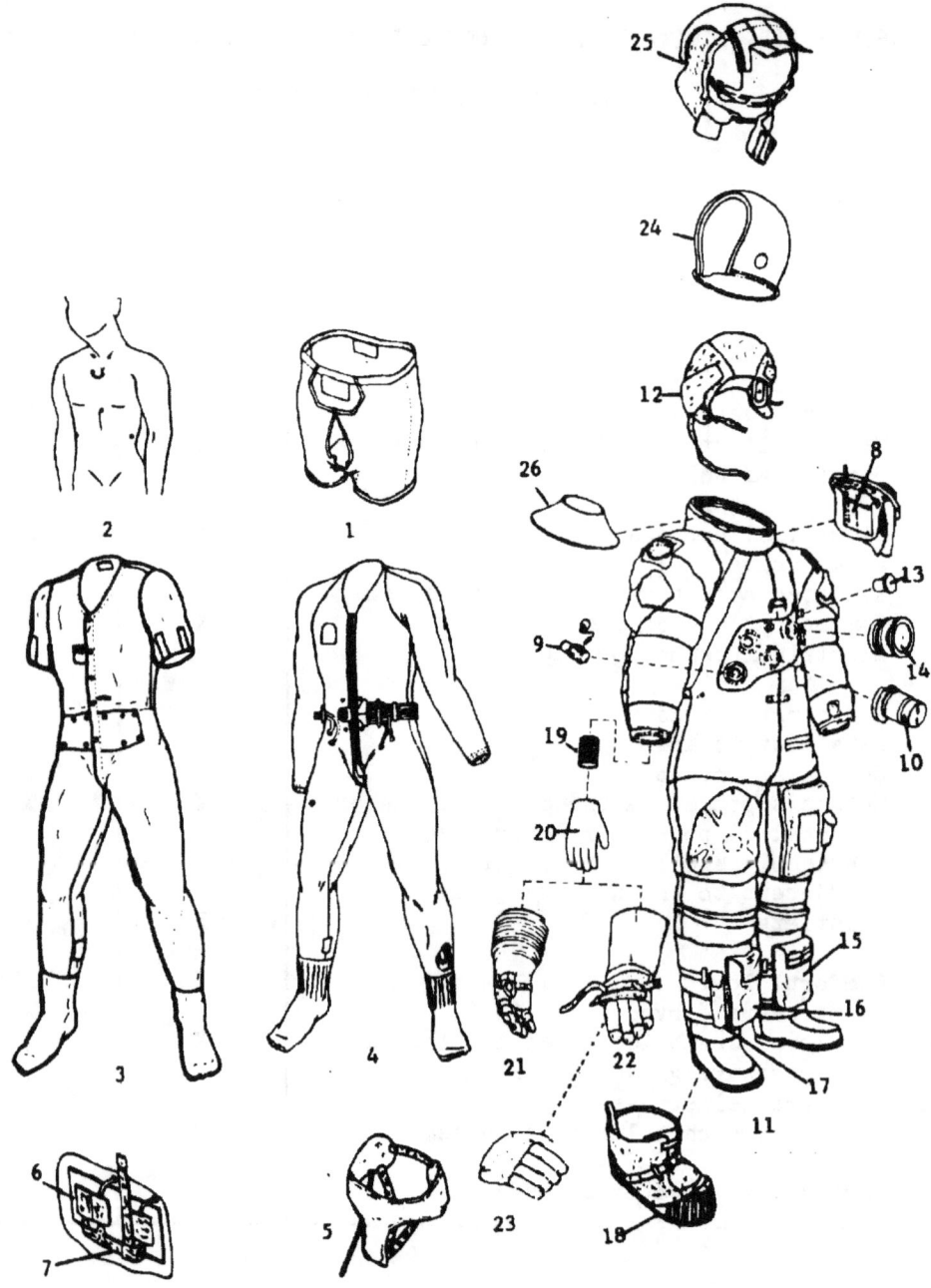

Figure 2-3.- EV A7LB pressure garment assembly and accessories interface configurations.

TABLE 2-IV.- EV A7LB PRESSURE GARMENT ASSEMBLY AND ACCESSORIES

INTERFACE CONFIGURATIONS

Components	Use	
	EV	IV
1. Fecal containment subsystem	X	X
2. Biomedical sensors	X	X
3. Constant wear garment (CWG)		X
4. Liquid cooling garment (used in place of CWG for EV and IV LM use)	X	
5. Urine collection and transfer assembly	X	X
6. Biomedical belt	X	X
7. Biomedical harness	X	X
8. Insuit drinking device	X	
9. Purge valve	X	
10. LCG receptacle plug		X
11. EV integrated torso limb suit assembly	X	X
12. Communications carrier	X	X
13. Electrical connector cap		X
14. Gas connector caps		X
15. Data list pocket (used as an EV contingency sample pocket)	X	X
16. Checklist pocket		X
17. Scissors pocket (attached to straps of checklist pocket or ITMG shell outboard of and adjacent to the utility pocket)	X	X
18. Lunar boots	X	
19. Wristlets	X	X
20. Comfort gloves	X	X
21. IV pressure gloves		X
22. EV glove assemblies (used in place of IV pressure gloves for EV use)	X	
23. Abrasion cover gloves (integrated with EV glove at pre-installation acceptance testing and used to protect the EV glove)	X	
24. Pressure helmet assembly	X	
25. Lunar extravehicular visor assembly	X	
26. Neck dam (for water egress)		

Amendment
11/5/71

The breathable gas used for respiration, pressurization, and ventilation is distributed within the pressurizable portion of the PGA through noncrushable ducts. Inlet and outlet connectors provide the interface between the suit ventilation distribution system and the spacecraft or PLSS environmental control system. A diverter valve (DV) directs the inlet gas flow to the helmet duct or diverts a portion of that flow to the torso duct as preferred by the crewman. The ventilating gas flows from the helmet down and over the body to the arm and leg extremities to remove body gas perspiration and heat. Outlet gas flows from the extremities through ducts to the exhaust connector. To preclude an accidental gas loss, a gas connector cap is provided for the unused connector port to prevent inadvertently depressing the poppet-type valve.

A manually operated purge valve may be fitted into the outlet gas connector. The purge valve is a part of the open-loop gas system that permits breathable gas from the oxygen purge system to flow through the PGA during emergency modes of pressurized suit operation.

An integrated thermal micrometeoroid garment (ITMG) is part of the EV torso limb suit. The assembly is a lightweight multilaminate unit designed to cover and conform to the contours of the torso limb suit assembly (TLSA). The cross section of materials for the ITMG affords protection against abrasion, thermal, and micrometeoroid hazards expected during free-space and lunar excursions. The outer layer is employed as a scuff and flame-impingement protective surface.

A receptacle on EV A7LB pressure garments connects the PLSS liquid cooling system to the liquid cooling garment (LCG) worn under the torso limb suit during EV excursions. The liquid cooling system removes metabolic heat from within the PGA. A plug is inserted into the multiple water connector receptacle when the LCG is not worn to preclude gas leakage from the pressurizable portion of the PGA.

A food and water port is provided in the side of the face area of the pressure helmet for emergency feeding and drinking.

Communications and biomedical data are transmitted through a suit electrical harness. The harness connector is mounted to the torso and provides an interface with the spacecraft or PLSS.

Biomedical instrumentation components employed within the PGA include electrocardiogram (ECG) and impedance pneumogram (ZPN) sensors that supply data to signal conditioners containe in a biomedical belt assembly, and a biomedical harness that provides an electrical interface between the signal conditioners and the suit electrical harness. The biomedical belt is snapped in place on the constant wear garment (CWG) or LCG.

The cotton fabric CWG is worn under the PGA next to the crewman's skin. The garment provides chafe protection and body cooling by perspiration wicking and evaporation. The CWG is worn as a comfort and cooling garment during IV modes of spacecraft operation.

The LCG replaces the CWG for lunar exploratory missions. The network of Tygon tubing within the LCG interfaces with the TLSA and PLSS to circulate water through the tubing network and transport metabolic heat from within the PGA.

To provide for emergency waste management, a fecal containment subsystem (FCS) is worn about the waist of the crewman next to the body for collecting and containing solid waste matter. A urine collection and transfer assembly (UCTA) collects waste liquids and provides an interface with the torso limb suit for transferring liquid from the UCTA to the spacecraft waste system.

The lunar extravehicular visor assembly (LEVA) fits over the pressure helmet to provide light and heat attenuation and to protect the crewman's eyes from harmful radiation during EV excursions.

A pair (one left and one right) of detachable EV glove abrasion covers fabricated from silicone-coated Nomex is integrated with the EV glove during preinstallation acceptance testing and permits handling of a core sample drill without damaging the EV gloves. The cover is installed over the EV glove with the access flap of the glove routed through the slot in the knuckle area of the cover. The Velcro hook patches inside the rear edge of the cover slot are engaged to the pile patches on the outside of the abrasion cover slot. The strap near the wrist area of the abrasion cover is engage to the Velcro hook attachment point to secure the cover over the EV glove. The abrasion covers may be readily removed after the drilling operation.

An insuit drinking device is mounted between the TLSA liner and inner pressure wall and contains drinking water for the crewman while performing lunar surface activities.

Pockets are available as a part of the PGA for stowage of miscellaneous flight articles. Penlight and pencil pockets are located on the left-shoulder and left-thigh areas. A sunglasses pocket is provided on the right shoulder. For storage of large items, a utility pocket is attached to the left thigh of the ITMG. Detachable checklist and data list pockets may be located below the knee of either leg or about the thigh of the left leg over the utility pocket. A scissors pocket is sewn to the straps of the detachable checklist pocket or secured to the ITMG shell outboard of and adjacent to the utility pocket.

To accommodate stowage of the equipment, provide for inflight maintenance, and protect equipment during an Apollo mission, the following flight support accessories are provided: an inflight helmet stowage bag (IHSB) for storing the LEVA, IV gloves, or EV gloves; an EMU maintenance kit that provides a lubricant for seals and "O" rings, helmet LEVA visors cleaning pads, replacement seals and emergency repair patches for the PGA; a helmet shield that fits over the PHA for scuff and abrasion protection during tunnel transfer; an inflight HSB for stowage and protection of the helmet shield and/or PHA; and an LCG adapter interconnect for connecting the LCG and the LM liquid cooling system during in LM rest periods with the PGA removed.

2.3.1 **EV A7LB Pressure Garment Assembly**

The EV A7LB PGA functions as a part of the EMU and the spacecraft environmental control system. The PGA is worn by the crew commander and LM pilot. The PGA contains a habitable environment and protects the astronaut from exposure to thermal and micrometeoroid conditions while he performs EV activites on the lunar surface or in free space.

The components comprising the PGA include:

a. EV A7LB TLSA
b. Pressure helmet assembly
c. Wristlets
d. Comfort gloves
e. IV pressure gloves
f. EV gloves
g. Data list pocket
h. Checklist pocket
i. Scissors pocket
j. Biomedical harness
k. Lunar boots
l. Neck dam

2.3.1.1 EV A7LB Integrated Torso Limb Suit Assembly

The EV ITLSA is a restrained, gas-retaining bladder structure integrated with a thermal micrometeoroid protective assembly and encompasses the crewman exclusive of the head and hands. The PHA and EV or IV pressure gloves are mated with the EV TLSA to complete a pressurizable envelope that protects the crewman in a depressurized spacecraft, free space, or the lunar environment. The assembly is composed of the following subassemblies as numbered in figure 2-4.

1. Gas connectors
2. Diverter valve
3. PLSS attachment (upper)
4. Outer electrical flange
5. Suit electrical harness
6. Multiple water connector
7. PLSS attachment (lower)
8. Pressure gage
9. Pressure gage cover
10. Liner
11. Ventilation ducts
12. Torso
13. Upper arms (r.h. and l.h.)
14. Pressure sealing slide fastener lock
15. Restraint cables
16. Boots (l.h. and r.h.)
17. Pressure relief valve
18. Legs (l.h. and r.h.)
19. Lower arms (r.h. and l.h.)
20. Restraint lock slide fastener
21. ITMG boots (l.h. and r.h.)
22. ITMG urine collection device clamp
23. ITMG arms (l.h. and r.h.)
24. ITMG torso
25. Water connector mounting ring
26. Core yarn and wrist ring
27. Lacing cord

The torso, upper and lower arms, legs, boots, and restraint cables are integrated to form the TLSA pressurizable vessel. This vessel includes convoluted joints which permit low-torque body movements and a near-constant-volume gas displacement within the PGA during normal joint flexure. Longitudinal cables extend across each convolute and sustain the axial loads. The neck, waist, shoulder cone, and ankle convolutes are of the constricted-restraint type, and the shoulder, elbow, knee, waist, and thigh joints are single-walled, integrated-restraint-and-bladder, bellows-like structures.

A textured nylon fabric is bonded to the inner surface of the pressure vessel to protect the bladder from scuffs, abrasions and snags.

An inner comfort liner within the TLSA is removable for cleaning and inspection. The assembly offers scuff protection to the wearer and covers the ventilation ducting to preclude accidental damage during suit-donning operations.

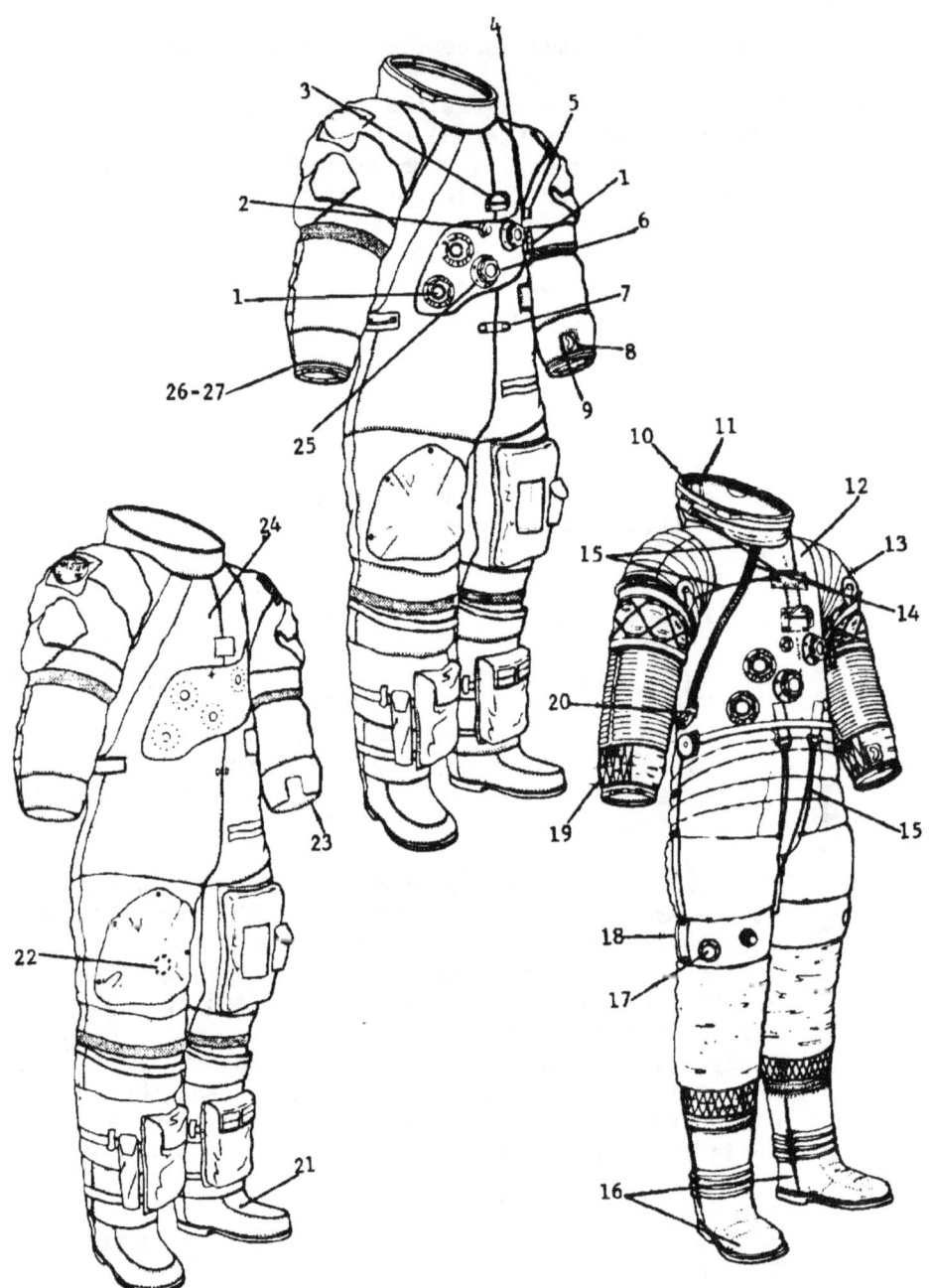

Figure 2-4.- EV A7LB integrated torso limb suit assembly.

Entry into the TLSA is made through restraint-and-pressure slide fasteners mounted in the waist area of the torso restraint-and-bladder layers. To preclude accidental opening lock assemblies are provided to hold the slide fasteners closed.

A network of noncrushable ducting laced to the inner TLSA surface, two sets of inlet and exhaust gas connectors, and a diverter valve comprise the ventilation distribution system within the TLSA. The TLSA and the ventilation distribution system interface with the pressure gloves and helmet to complete the PGA pressurization and ventilation system.

A pressure gage is mounted on the left-arm wrist cone, and a pressure relief valve is mounted on the right-leg thigh cone. The pressure gage indicates differential pressures of from 2.5 to 6.0 psid, and the pressure relief valve relieves pressures in excess of 5.0 psid.

The suit electrical harness provides signal paths for biomedical instrumentation data and communications transmissions. The suit-mounted connector permits an electrical and mechanical interface with the spacecraft or PLSS communications umbilical.

A flange-mounted multiple water connector secured to the torso provides a mechanical mate between the LCG and PLSS or LM liquid cooling systems. When the LCG is not worn, a plug is locked into the connector opening to provide a gas seal.

The ITMG torso, arms, boots, and pressure gage cover afford flame impingement, thermal, and micrometeoroid protection to the pressurizable portion of the TLSA and to the crewman. The assemblies employ a multilayered cross section as shown in table 2-V.

The water connector mounting ring, outer electrical flange, ITMG urine collection device (UCD) clamp, core yarn, wrist ring, and lacing cord secure the thermal and micrometeoroid protective assemblies to the torso limb suit.

Amendment 2
11/5/71

TABLE 2-V.- EV A7LB ITMG MATERIALS CROSS SECTION

(LISTED FROM THE INSIDE OUT)

Nomenclature	Function
Rubber-coated nylon (ripstop)	Inner liner and micrometeoroid protection
Nonwoven Dacron	Thermal spacer layer
Aluminized Mylar film	Thermal radiation protection
Nonwoven Dacron	Thermal spacer layer
Aluminized Mylar film	Thermal radiation protection
Nonwoven Dacron	Thermal spacer layer
Aluminized Mylar film	Thermal radiation protection
Nonwoven Dacron	Thermal spacer layer
Aluminized Mylar film	Thermal radiation protection
Nonwoven Dacron	Thermal spacer layer
Aluminized Mylar film	Thermal radiation protection
Beta marquisette	Thermal spacer layer
Gridded[a] aluminized Kapton film	Thermal radiation protection
Beta marquisette	Thermal spacer layer
Gridded aluminized Kapton film	Thermal radiation protection
Beta marquisette	Thermal spacer layer
Teflon-coated yarn Beta cloth	Flame impingement layer
Teflon fabric	Abrasion layer

[a] A 2-inch gridding with Polyemite tape is employed in the arm and knee areas; 4-inch gridding is provided in all other areas.

2.3.1.1.1 **EV A7LB torso limb suit assembly.**- The TLSA is sized to fit a specific crewman. To further customize the fit of a torso limb suit, to optimize mobility in the suit, and to provide maximum comfort, the following adjustments may be made.

    a. Neck height   e. Arm length
    b. Neck angle   f. Crotch height
    c. Shoulder width   g. Crotch and limb angle
    d. Elbow convolute height   h. Leg length

The torso section and shoulder, wrist, thigh, and lower leg cones employ a bilayered cross section, an inner gas retention layer, and an outer structural restraint layer to maintain the optimum shape and size of the torso limb suit during pressurized and depressurized modes of suit operation. The inner bladder layer is loosely fitted to the restraint layer and is attached to the restraint layer at strategic points for support and alinement. The convolutes provided at the shoulder, elbow, thigh, and knee areas are flexible, single-walled structures or joints to satisfy suit mobility requirements. Movements in the joint areas cause little change in the volume of gas within the PGA, but displace the gas within the joint area.

The TLSA boot assembly includes an outer fabric restraint, a sole and heel assembly, and an inner rubber bladder. The heel and sole assemblies employ an inner core of aluminum honeycomb in the heel and arch areas and a stainless steel truss core in the front sole area. The areas where honeycomb is used are rigid, and the truss area permits longitudinal flexibility to accommodate normal foot movements.

Nylon webbings at cable attachment points evenly distribute restraint loads. Metal eyelets and grommets line and reinforce the holes provided for cable attachment points.

An abrasion layer secured to the inner bladder wall reduces wear normally caused by direct contact between the body and the bladder.

Noncollapsible ducts along the inner wall of the TLSA make up the ventilation distribution system. Each duct is constructed of parallel lengths of nylon spacer coils wrapped with a nylon mesh cloth. The nylon mesh cloth and spacer construction are dipped in a rubber compound which promotes rigidity of the cloth and adds a nonslip characteristic between the cloth and the coil spacers. The assembled unit is then wrapped with bladder material to form a noncrushable duct with an airtight wall. These ducts are secured to the TLSA by a system of loop-type and lacing cord.

A comfort liner in the interior of the TLSA facilitates donning and promotes comfort. The leg of the liner assembly is zipped to the boot liner at the lower leg area. The assembly is secured to the torso limb suit with hook and pile fastener tape and snap fasteners at the neck opening, around the wrists, and along each side of the entry closure. Synthetic elastomer foam pads over each shoulder and at the biceps area of the arm promote comfort. Reinforced openings through the liner provide passages for the suit electrical harness communications branch, biomedical instrumentation branch, and urine transfer hose. A communications snap-flap at the front of the neck opening holds the communications branch in place to facilitate donning. The front-knee panels and the rear-elbow panels of the liner are pleated along each side to form semipockets which afford relief during limb flexation.

2.3.1.1.2 Lunar integrated thermal micrometeoroid garment.- The ITMG is sized to fit and conforms to the contours of the torso limb suit. The ITMG may be removed from the torso limb suit for inspection and maintenance. The multilaminate cross section of the ITMG prevents thermal damage to, and punctures in, the torso limb suit, and protects the crewman from the extreme temperatures and micrometeoroid flux densities normally expected on the lunar surface and in the free space within the lunar orbit perimeter. To protect against fire and exposed surface abrasion, an outer layer of Teflon fabric and an inner layer of Teflon-coated yarn Beta cloth are provided.

For protection from the thermal environment of free space and the moon, seven layers of aluminized film are used to reflect radiant heat and to reduce heat conduction between the aluminized film layers. A low-heat-conducting fabric of nonwoven Dacron or Beta marquisette is used to separate each layer of film. An inner layer of ripstop fabric, the thermal protective layers, and the fire impingement and abrasion layers provide the mass needed to afford micrometeoroid protection to the TLSA and crewman.

2.3.1.2 Pressure Helmet Assembly

The PHA (fig. 2-5) is a transparent bubble which engages with the torso limb suit and encloses the crewman's head. The assembly consists of an anodized aluminum neck ring, a vent pad, a valsalva device, and a transparent polycarbonate shell.

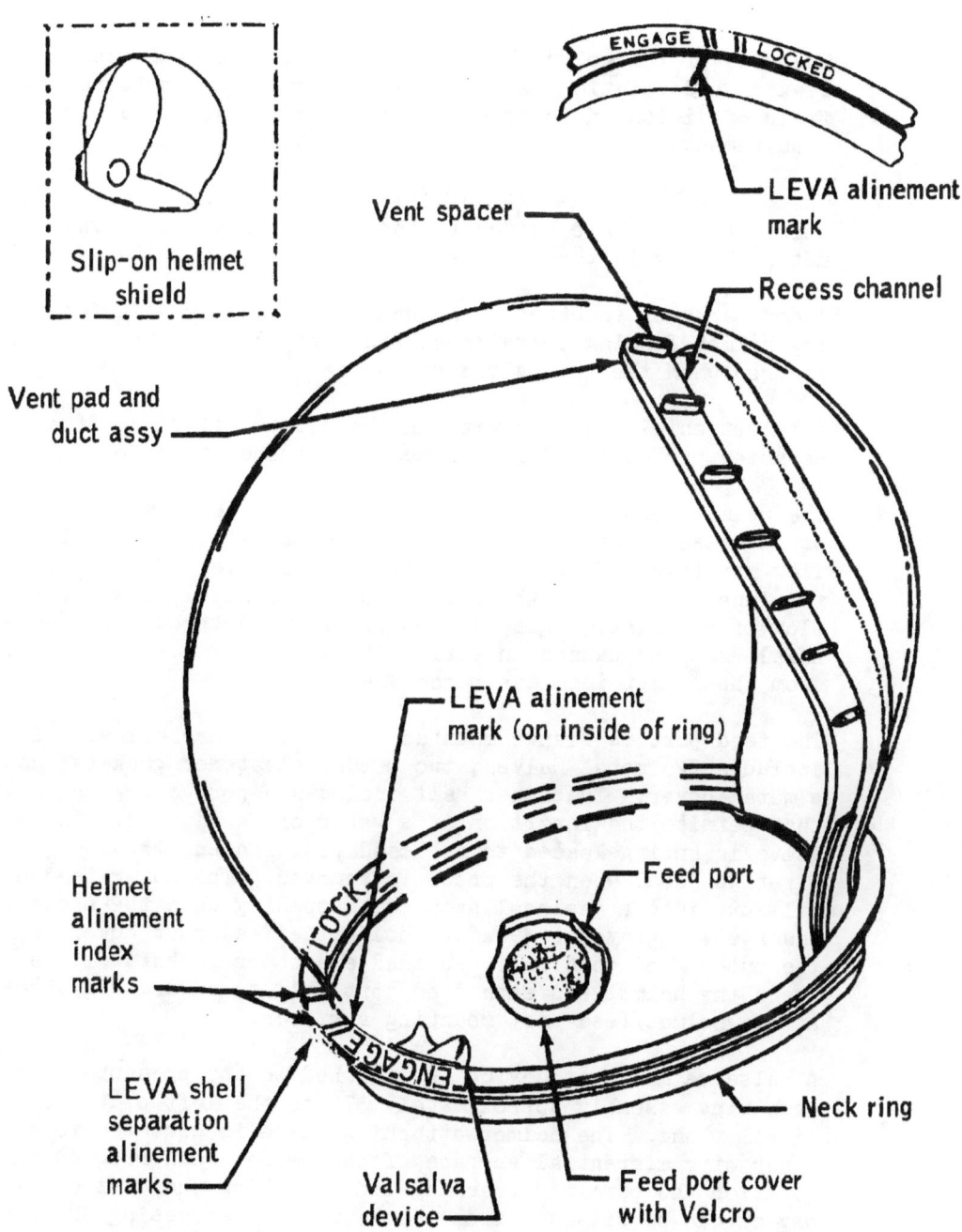

Figure 2-5.- Pressure helmet assembly and helmet shield.

The size of the polycarbonate shell permits normal neck flexion and rotation movements and provides an unobstructed field of vision in accordance with specified optical requirements.

The polycarbonate helmet shell is molded and has a machined bayonet base bonded to the helmet neck ring. The helmet neck ring is the male half of the suit neck ring assembly.

Index marks on each neck ring half are used for alinement during helmet donning operations, and a rigid airtight joint is assured when the two halves are joined.

A helmet shield is used with the helmet to afford scuff and abrasion protection during spacecraft tunnel transfers.

The helmet vent pad bonded to the back of the helmet shell provides shock protection and is used as a helmet ventilation flow manifold. Vent pad louvers guide a layer of gas along the inner surface of the helmet to the oronasal area. This flow of ventilation gas is then distributed through the oronasal area and causes an efficient exhaust of carbon dioxide from the helmet into the torso area.

The feed port is flange mounted to the pressure helmet and includes two metal halves, two beaded elastomer gaskets, and a metal cover. The inner half includes a port and gate valve that permits the insertion of a water or food probe. The valve is spring loaded to a closed position and provides an air-tight seal when the probe is removed. The outer feed port half provides a gas seal around the opening when the probe is inserted. A bayonet juncture holds the feed port cover to the outer feed port half. Beaded elastomer gaskets fit between the helmet and each feed port half to ensure a gas seal at the helmet/feed port mounting surfaces.

A valsalva maneuver device is attached to the pressure helmet neck ring assembly approximately 37° to the left of the sagittal plane. The helmet attaching plate is cemented to the inner circumferential surface of the helmet neck ring at this location and permits attaching and detaching the device. The device can be detached from the helmet by depressing the latch and sliding the device in either direction until free of the helmet attaching plate.

2.3.1.3   Wristlets

The wristlets (fig. 2-6) may or may not be selected for use by the crewman for comfort. The wristlets are cylindrically shaped and constructed of ribknit cotton material. The wristlets may be attached to the comfort glove to provide the wrist and lower arm with protection against wrist disconnect buffeting.

2.3.1.4   Comfort Gloves

The comfort gloves (fig. 2-6) may be used by the crewman for comfort. When used, they are worn beneath the EV or IV PGA pressure gloves to avoid chafe between the skin and the gloves. The comfort gloves are made of nylon tricot material and are available in either long or short lengths of standard small, medium, or large sizes. The long-length gloves are also available in custom sizes.

2.3.1.5   IV Pressure Glove Assembly

The pressure glove assembly (fig. 2-6) is a flexible, gas-retaining device which locks to the torso limb suit by means of a quick-disconnect coupling (the wrist disconnect). The bladder assembly is dip molded from a hand cast of the individual's hand. The bladder is comprised of an inner restraint core of nylon tricot covered with a dipped rubber compound. The dexterity of the bladder is increased by built-in relief projections over the knuckle areas, and, to facilitate thumb extension, a gusset is provided in the thumb/forefinger crotch

A standard convolute section is incorporated in the wrist area of the bladder to allow omnidirectional movement of the wrist. The convoluted section is restrained by a nylon restraint fabr layer and a system of sliding cables secured to a wrist restra ring and the glove side-wrist disconnect. The cable restraint system accepts the axial load across the glove convolute.

The glove side-wrist disconnect is the male portion of the wrist disconnect assembly and features a sealed bearing which permits 360° glove rotation.

The fingerless glove/outer convolute cover is a restraint assembly which is cemented onto the bladder at the wrist area and encloses the entire hand and wrist exclusive of the fingers and thumb. An external palm restraint assembly minimizes the ballooning effect when pressurized, thereby enhancing grip control. The convolute covers protect the bladder and convolute restraint system.

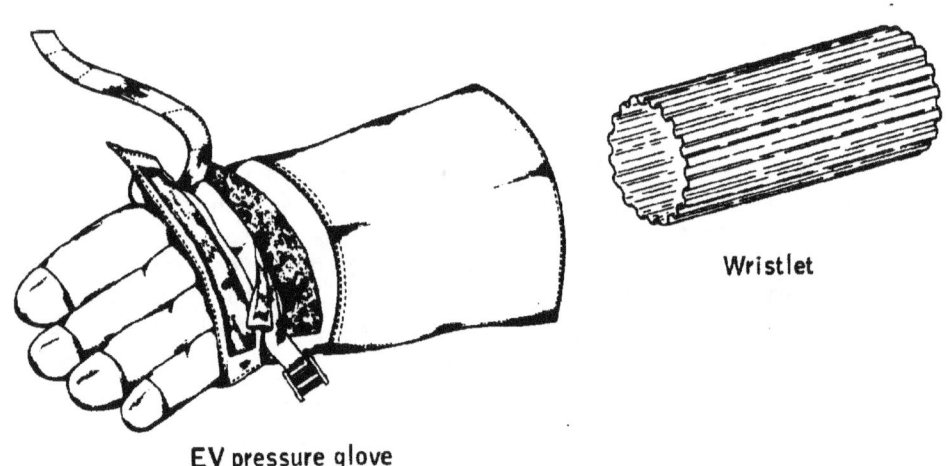
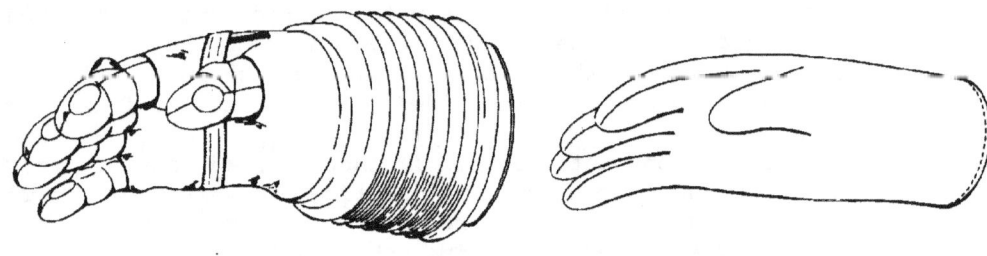

Figure 2-6.- Glove assemblies with wristlets.

2.3.1.6   EV Pressure Gloves

The EV glove assembly (fig. 2-6) is a protective hand covering interfaced with the torso limb suit assembly prior to egress for extravehicular operations. The EV glove consists of a modified IV pressure glove assembly covered by the EV glove shell assembly. The assembly covers the entire hand and has an integral cuff or gauntlet which extends the protective covering well above the wrist disconnect.

A lunar surface EVA checklist is attached to the EV glove gauntlet outer shell as a crew/mission requirement. The specific location, method of attachment, and orientation of the checklist on the glove gauntlet will be defined by the crewman to satisfy specific needs and mission objectives.

The EV glove thermal shell is a multilayered assembly (table 2-VI) which provides scuff, abrasion, flame impingement, and thermal protection to the pressure glove and crewman. A woven metal (Chromel R) fabric is incorporated over the hand area for added protection from abrasion. The thumb and finger shells are made of high-strength silicone rubber which is reinforced with nylon cloth and provides improved tactility and strength. A silicone dispersion coating is applied to the palm, around the thumb, and to the inner side of each finger for increased gripping.

The outer cover is shaped to the inner pressure glove and does not appreciably restrict the dexterity of the inner pressure glove. A flap is sewn onto the back of the glove shell and provides access to the palm restraint flap. The flap is opened or closed by engaging or disengaging the hook-and-pile fastener tape. When the palm restraint flap and hook-and-pile tapes are disengaged, the glove shell can be removed by disengaging the cemented interfacing areas near the fingertips. The materials cross section of the cover layer of the EV glove assembly is identified in table 2-VI.

TABLE 2-VI.- MATERIALS CROSS SECTION FOR EV THERMAL GLOVE

Material	Function
Pressure glove	Pressure retention
Aluminized Mylar (7 layers)	Insulation film
Nonwoven Dacron (6 layers)	Insulation spacer
Teflon-coated Beta yarn (gauntlet only)	Fire resistant shell (gauntlet only)
Teflon cloth (gauntlet only)	Abrasion resistant (gauntlet only)
Chromel R metal fabric (hand only)	Abrasion, fire, heat resistant
Silicone rubber (finger tips only)	Increase friction.

2.3.1.7   Data List Pocket Assembly

The data list pocket assembly (fig. 2-7) is a strap-on assembly which is normally wrapped around the lower left or right leg of the ITMG. The pocket is attached to the leg by two straps held in place by belt loops. The pocket opens and closes by means of an overhanging flap secured by strips of hook-and-pile fastener tape. The data list pocket may be provided as an EV contingency sample pocket. The walls of the pocket include removable stiffeners which hold the pocket open to reduce interferences while inserting or removing articles.

The pocket may be secured to the left thigh in an upright or upside down attitude to attain maximum accessibility to the pocket. Hook-and-pile fastener tape is employed to hold the pocket flap in the open position when the pocket is upright and secured to the thigh.

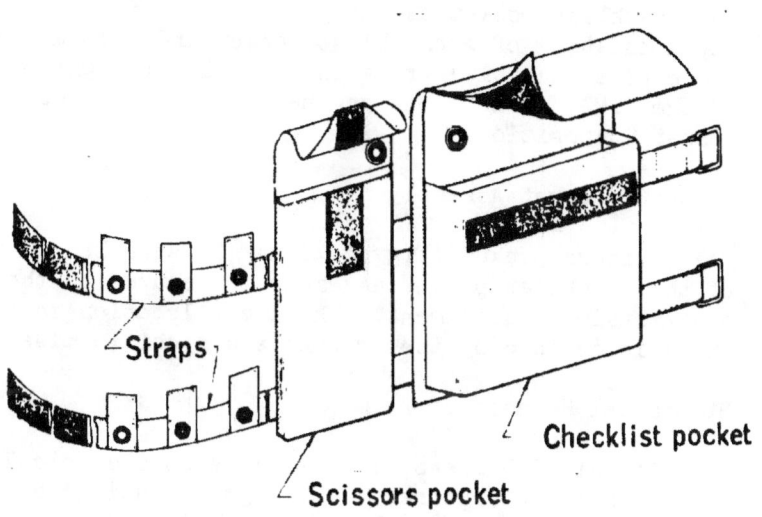
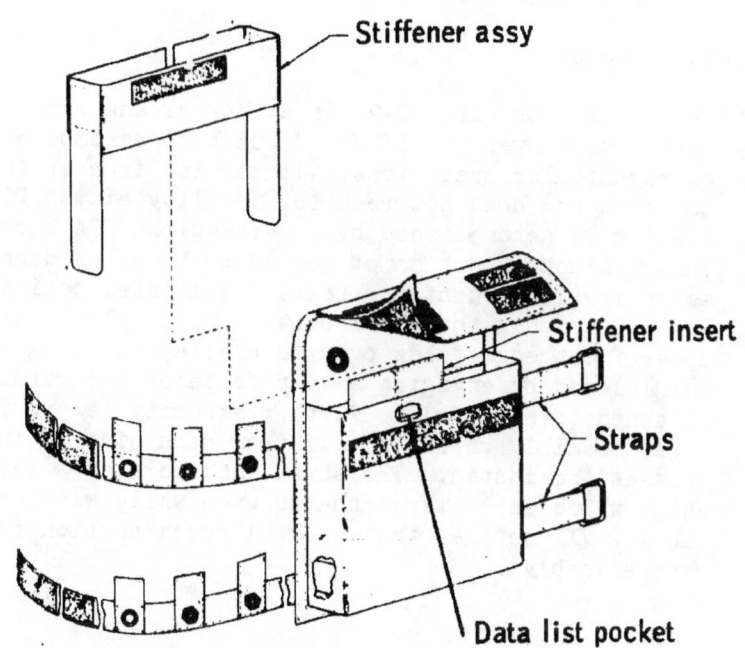

Figure 2-7.- Detachable pocket assemblies.

2.3.1.8   Checklist Pocket Assembly

The checklist pocket assembly (fig. 2-7) is a strap-on assembly consisting of a checklist pocket and belt assemblies. The entire assembly straps onto the lower right or left leg of the ITMG. Belt loops on the legs of the ITMG hold the pocket in position.

2.3.1.9   Scissors Pocket Assembly

The scissors pocket (fig. 2-7) may be attached to the straps of the checklist pocket assembly or secured to the ITMG as a crew/mission requirement. The exact location on the ITMG shell is defined by the crewman and specific mission objectives.

2.3.1.10   Biomedical Harness

The biomedical harness (fig. 2-8) is an electrical cable assembly which interconnects the signal conditioners and dc-to-dc converter within the biomedical belt and interfaces with the suit electrical harness.

2.3.1.11   Lunar Boots

The lunar boot (fig. 2-9) is a thermal and abrasion protective device worn over the ITMG and PGA boot assemblies during lunar extravehicular operations. It permits free articulation of the foot and does not restrict mobility of the PGA boot. Donning is accomplished by inserting the PGA boot into the enlarged upper portion of the lunar boot. A donning strap assembly (located at top rear) facilitates positioning of the PGA boot within the lunar boot. The surplus material at the upper front edge folds over to overlap the tongue area and is held closed by engaging a snap fastener and retaining strap attached to each fold. Further security is provided by a strap assembly which extends from each side of the heel and crosses the instep. The strap incorporates a latching mechanism which is easily actuated even while wearing EV gloves. Table 2-VII defines the material cross section of the lunar boot assembly.

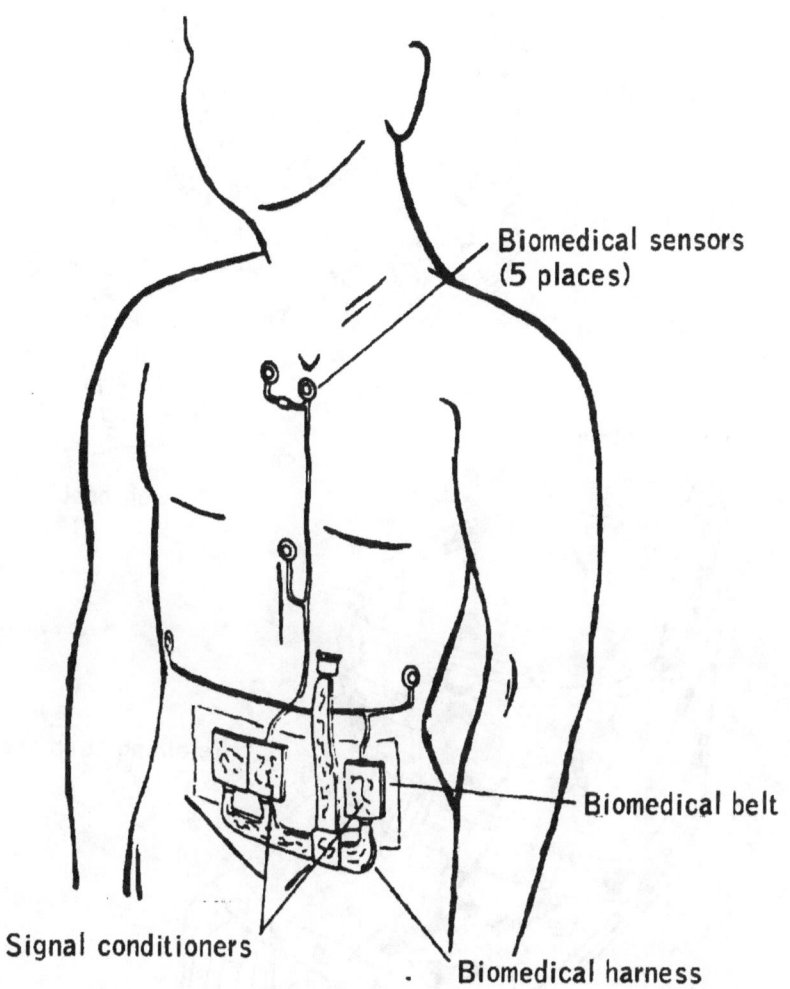

Figure 2-8.- Biomedical harness and sensors.

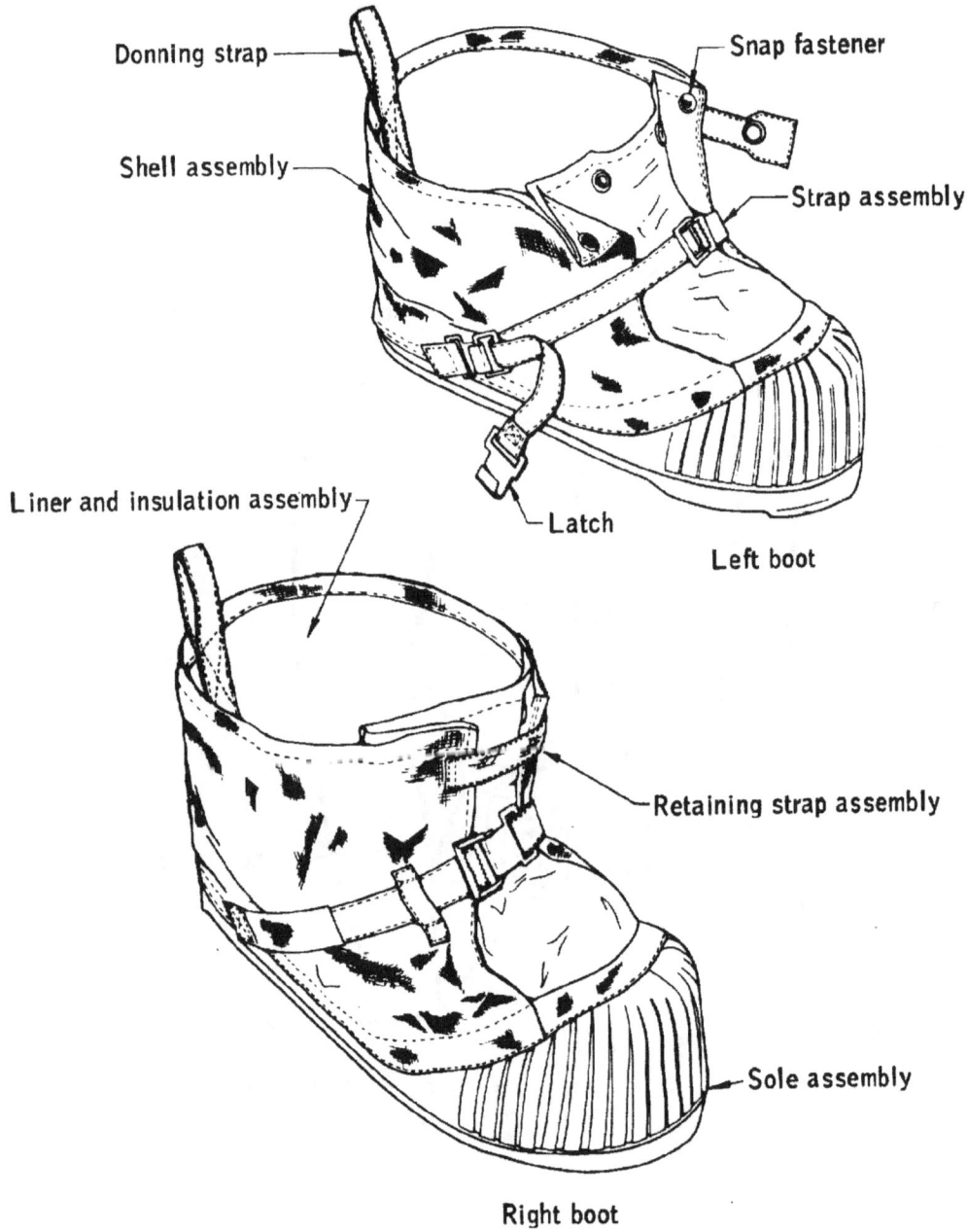

Figure 2-9.- Lunar boots.

TABLE 2-VII.- MATERIALS CROSS SECTION FOR LUNAR BOOT

Material	Function
Teflon-coated Beta cloth	Boot liner
Aluminized Mylar	Insulation film
Nomex felt	Thermal boot pad
Aluminized Mylar (9 layers)	Insulation film
Nonwoven Dacron (9 layers)	Insulation spacer
Beta marquisette Kapton laminate (2 layers)	Outer insulation
Teflon-coated Beta cloth	Fire resistant shell
High-strength silicone rubber	Lunar boot sole
Chromel R metal fabric	Abrasion, fire, heat resistant

2.3.1.12 Neck Dam

The neck dam assembly (fig. 2-10) is a sealing device to prevent water seepage into the TLSA through the neck opening during suited operations in the water. The assembly consists of a neck dam seal constructed of rubber, a neck dam ring assembly made of flexible metal, and a storage lanyard. The neck dam assembly is conically shaped with a sized opening for the head and neck. The neck dam is donned after reentry and just prior to spacecraft egress operations. The size of the neck dam is determined by the circumference of the head and neck opening in the neck dam seal. The size can be identified by the part number suffix (-1400, neck size 14; -1450, neck size 14-1/2; etc.), and it is available in sizes 13-1/2 to 16-1/2.

2.3.2 <u>CMP A7LB Pressure Garment Assembly</u>

The CMP A7LB PGA functions as a part of the spacecraft environmental control system or the EMU. The PGA contains a habitable environment and protects the astronaut from exposure

Size designation	Size, in.	Dim. A, in.
A7L-121036-02-1350	13-1/2	5-3/4
A7L-121036-02-1400	14	5-1/2
A7L-121036-02-1450	14-1/2	5-1/4
A7L-121036-02-1500	15	5
A7L-121036-02-1550	15-1/2	4-3/4
A7L-121036-02-1600	16	4-1/2
A7L-121036-02-1650	16-1/2	4-1/4

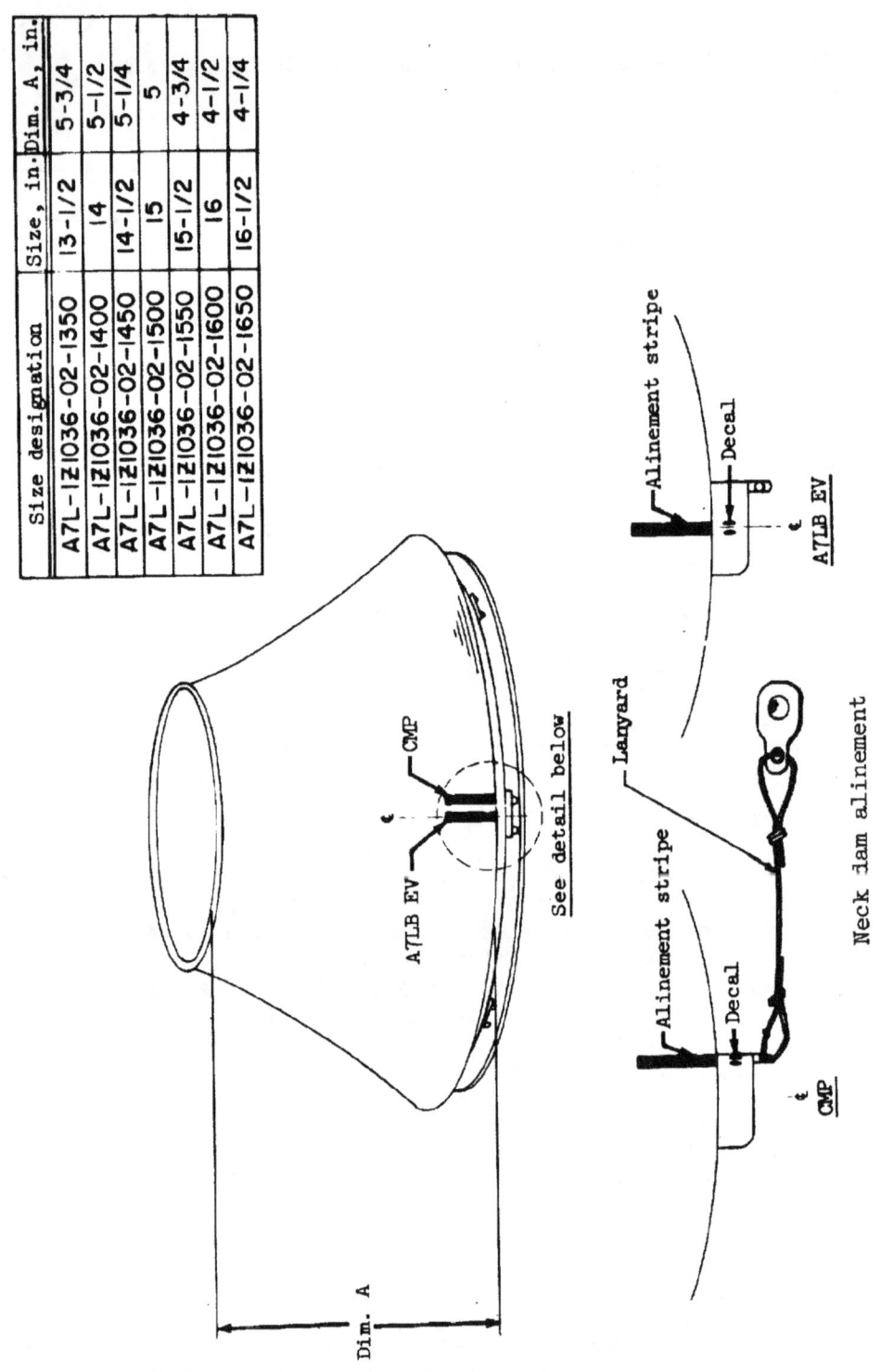

Figure 2-10.- Neck dam.

to thermal and micrometeoroid conditions during EV activities in the free space within the lunar orbit perimeter. The components of the PGA include:

a. CMP A7LB ITLSA
b. PHA
c. Wristlets
d. Comfort gloves
e. IV pressure gloves
f. EV gloves
g. Data list pocket
h. Checklist pocket
i. Scissors pocket
j. Biomedical harness
k. Neck dam

2.3.2.1 CMP A7LB Integrated Torso Limb Suit Assembly

The CMP ITLSA is a restrained, gas-retaining bladder structure integrated with a thermal micrometeoroid protective assembly. The CMP ITLSA encompasses the crewman exclusive of the head and hands. The PHA and EV or IV pressure gloves are mated with the CMP TLSA to complete a PGA for protecting the crewman in a depressurized spacecraft or free space environment. The ITLSA consists of the following subassemblies as numbered in figure 2-11.

1. Torso
2. Pressure gage
3. Torso adjusting strap
4. Restraint cables
5. Pressure sealing slide fastener
6. Boots (r.h. and l.h.)
7. Legs (r.h. and l.h.)
8. Pressure relief valve
9. Gas connectors with diverter valves
10. Arm assembly
11. Suit electrical harness
12. Upper arms (r.h. and l.h.)
13. Liner
14. Core yarn, wrist ring and lacing cord
15. Cover layer assembly boots (r.h. and l.h.)
16. UCD and medical injection access flap
17. Cover layer assembly arms (r.h. and l.h.)
18. Pressure gage cover
19. Cover layer assembly torso
20. Ventilation ducts (not shown)
21. Outer electrical flange (not shown)
22. ITMG UCD clamp (not shown)

The torso, upper and lower arms, legs, boots, and restraint cables are integrated to form the CMP TLSA pressurizable vessel. This vessel includes convoluted joints for low-torque body movements and a near-constant volume displacement during normal joint movements. Longitudinal cables extend across each convolute and sustain the axial loads. The shoulder cone and ankle convolutes are of the constricted-restraint type; and the shoulder, elbow, knee, waist, and thigh joints are single-walled, integrated restraint and bladder, bellows-like structures.

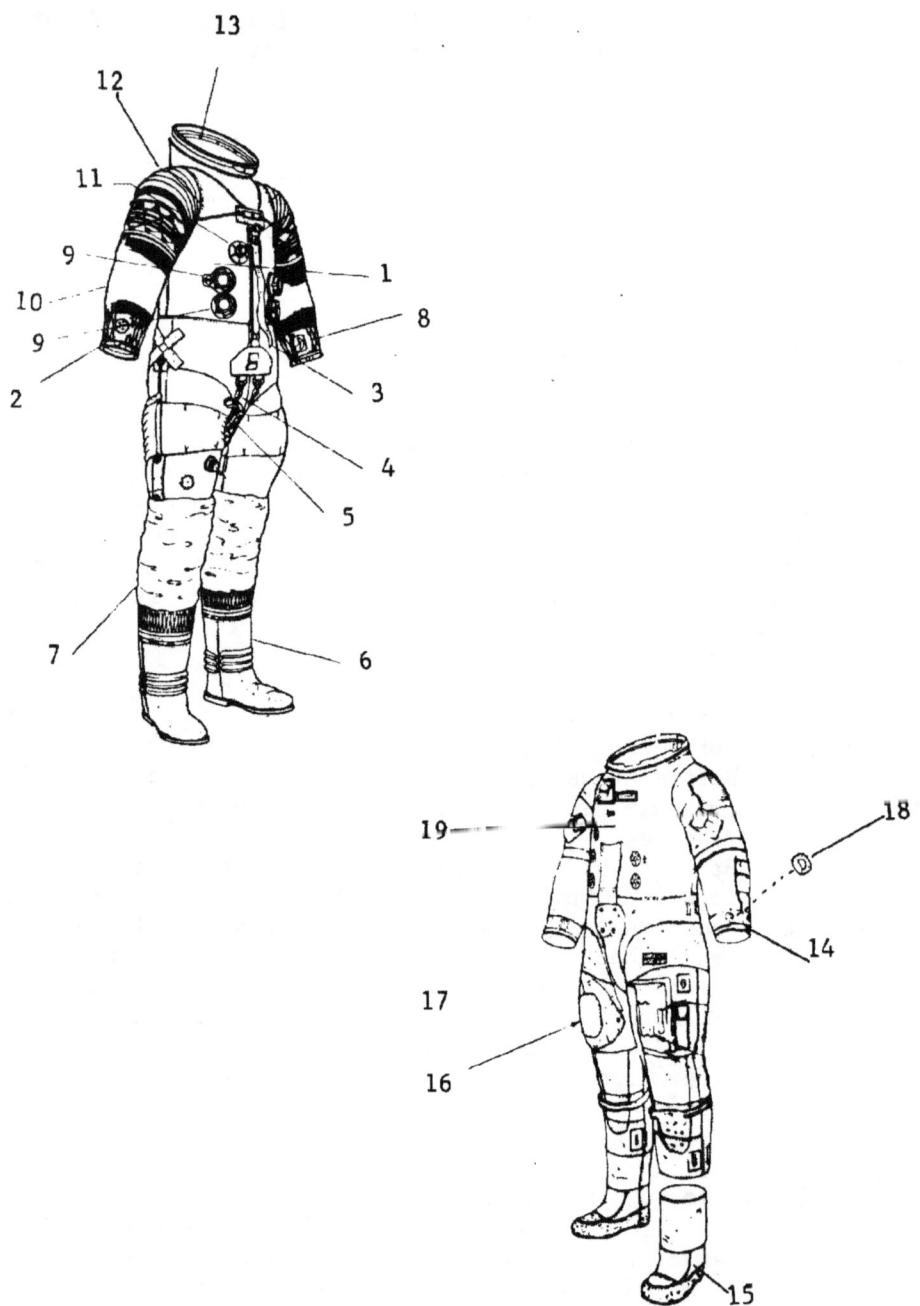

Figure 2-11.- CMP A7LB integrated torso limb suit.

An inner comfort liner within the TLSA is removable for cleaning and inspection. The assembly offers scuff protection to the wearer and covers the ventilation ducting to preclude accidental damage during suit-donning operations.

Entry into the TLSA is made through an integrated restraint and pressure slide fastener assembly mounted vertically along the spinal column and through the crotch area. To preclude accidental opening, a lock assembly for the pressure sealing slide fastener holds it in the closed position.

A network of noncrushable ducting secured to the inner TLSA surface, two sets of inlet and exhaust gas connectors, and a diverter valve for each inlet connector comprise the ventilation distribution system within the TLSA. The TLSA and a ventilation distribution system interface with the pressure gloves and helmet to complete the PGA pressurization and ventilation system. A pressure gage is mounted on the left-arm wrist cone, and a pressure relief valve is mounted on the left arm. The pressure gage indicates differencial pressures of from 2.5 to 6.0 psid, and the pressure relief valve reliev pressures in excess of 5.0 psid.

The suit electrical harness provides a signal path for biomedical instrumentation data and communications transmissions The suit-mounted connector permits an electrical and mechanical interface with the spacecraft or PLSS communications umbilical.

The cover layer assembly (CLA) torso, arms, boots, and pressure gage cover afford flame impingement, thermal, and micrometeoroid protection to the pressurizable portion of the TLSA and to the crewman. The assemblies employ a multilayered cross section as shown in table 2-VIII.

The outer electrical flange, ITMG UCD clamp, core yarn, wrist ring, and lacing cord secure the thermal and micrometeoroid protective assemblies to the torso limb suit.

Amendment
11/5/71

TABLE 2-VIII.- CMP A7LB CLA MATERIALS CROSS SECTION

(LISTED FROM THE INSIDE OUT)

Nomenclature	Function
Rubber-coated nylon (ripstop)	Inner liner
Aluminized Mylar film	Thermal radiation protection
Nonwoven Dacron	Thermal spacer layer
Aluminized Mylar film	Thermal radiation protection
Nonwoven Dacron	Thermal spacer layer
Aluminized Mylar film	Thermal radiation protection
Nonwoven Dacron	Thermal spacer layer
Aluminized Mylar film	Thermal radiation protection
Nonwoven Dacron	Thermal spacer layer
Aluminized Mylar film	Thermal radiation protection
Aluminized Kapton film/ Beta marquisette laminate	Fire and thermal radiation protection
Aluminized Kapton film/Beta marquisette laminate	Fire and thermal radiation protection
Teflon-coated yarn Beta cloth	Fire protection
Teflon fabric	Abrasion protection

2.3.2.1.1 **CMP A7LB torso limb suit assembly.**- The CMP TLSA is similar to the EV TLSA described in paragraph 2.3.1.1.1 except for the following details.

    a. The ventilation distribution system ducts are secured to the TLSA in the EV configuration by a system of loops and lacing cord and, in the CMP configuration, by hook and pile fastener tape and bonding strips.

b. The semipockets at the knees of the comfort liner are formed by front panel pleats in the EV configuration and by rear panel pleats in the CMP configuration.

2.3.2.1.2 <u>CMP cover layer assembly</u>.- The CLA is identical to the lunar ITMG described in table 2-VIII.

2.3.2.2 Pressure Helmet Assembly

The CMP PHA is identical to the EV PHA described in paragraph 2.3.1.2 and figure 2-5.

2.3.2.3 Wristlets

The CMP wristlets are identical to the EV wristlets described in paragraph 2.3.1.3 and figure 2-6.

2.3.2.4 Comfort Gloves

The CMP comfort gloves are identical to the EV comfort gloves described in paragraph 2.3.1.4 and figure 2-6.

2.3.2.5 IV Pressure Gloves

The CMP pressure glove assembly is identical in all respects to the EV pressure glove assembly described in paragraph 2.3.1 and figure 2-6.

2.3.2.6 EV Gloves

Refer to paragraph 2.3.1.6 and figure 2-6.

2.3.2.7 Data List Pocket

Refer to paragraph 2.3.1.7 and figure 2-7.

2.3.2.8 Checklist Pocket

Refer to paragraph 2.3.1.8 and figure 2-7.

2.3.2.9 Scissors Pocket

Refer to paragraph 2.3.1.9 and figure 2-7.

2.3.2.10 Biomedical Harness

Refer to paragraph 2.3.1.10 and figure 2-8.

2.3.2.11   Neck Dam

Refer to paragraph 2.3.1.12 and figure 2-10.

2.3.3   Interface Components

This paragraph contains descriptions of the components which interface the torso limb suit with other components of the EMU or with the spacecraft, and those which are provided as accessories to the suit. The interface and accessory components are as follows.

a. PLSS attachments
b. Tether attachments
c. Helmet attaching ring
d. Wrist disconnects
e. Gas connectors
f. Diverter valve
g. Multiple water connector
h. Urine transfer connector
i. Medical injection patch
j. Zipper lock assemblies
k. Pressure relief valve
l. Biomedical belt
m. Biomedical harness
n. Suit electrical harness

2.3.3.1   PLSS Attachments

Two attachment brackets (fig. 2-12) on the EV A7LB PGA anchor the shoulder and waist PLSS support straps in place. The upper bracket is fixed to the torso sternum area. The lower PLSS attachment is fitted over the ITMG and snapped to the front torso crotch cable "D" rings located in the abdominal area.

2.3.3.2   Tether Attachment

Tether attachments (fig. 2-13) are available at the left and right sides of the EV PGA. The attachment interfaces with and becomes a part of the LM tether system. The LM tether system with the PGA tether attachments provide an artificial gravity to assist the crewman in maintaining stability within the LM.

2.3.3.3   Helmet Attaching Ring Assembly

The helmet is attached to the TLSA by a self-latching, self-sealing, quick-disconnect coupling (fig. 2-14). The TLSA side of the coupling consists of a neckring housing, eight latch assemblies, a rotating locking ring, and a pushbutton lock subassembly on the locking ring. Index marks and

Amendment 2
11/5/71

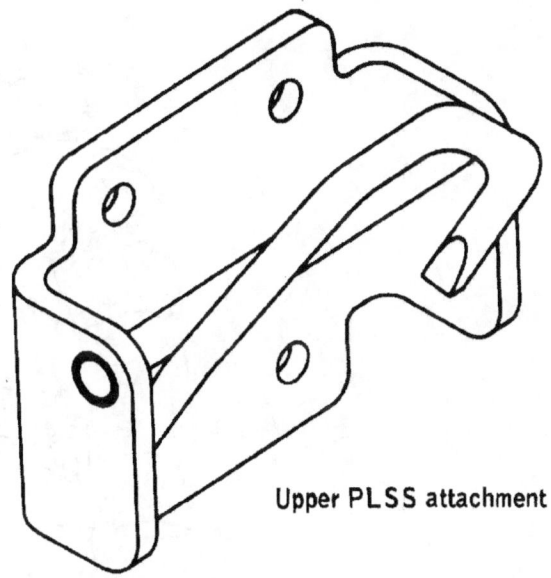

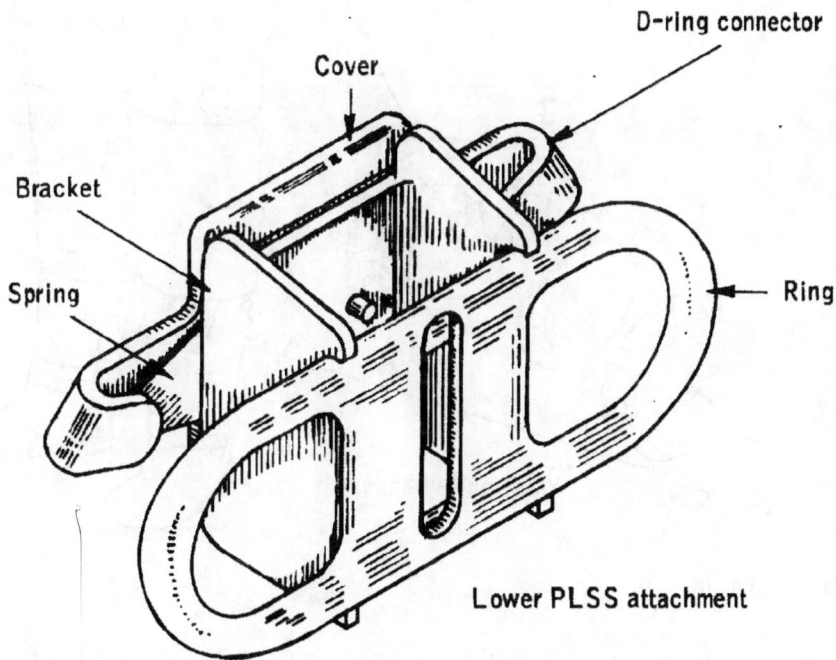

Figure 2-12.- PLSS attachments.

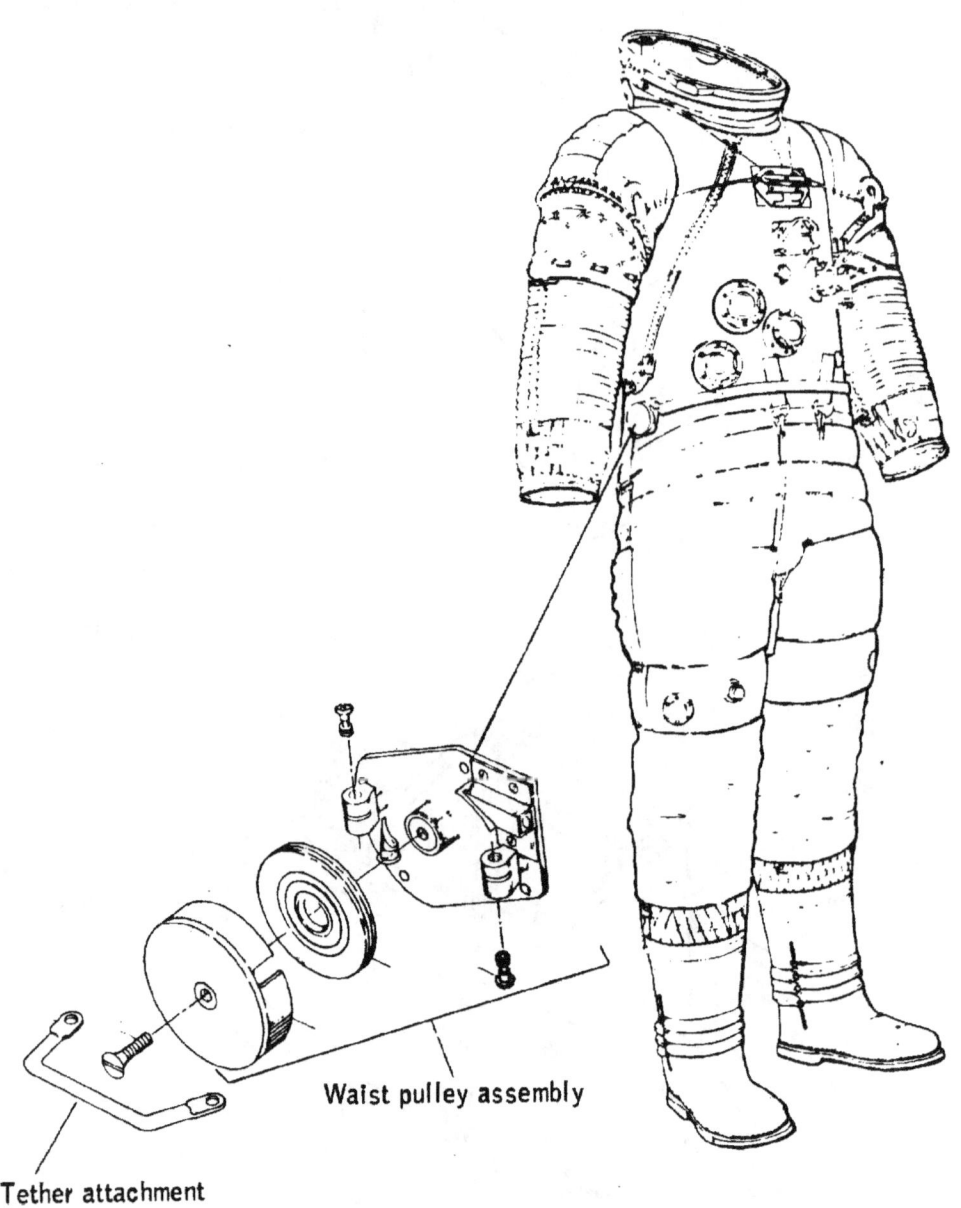

Figure 2-13.- Lunar module tether attachments (A7LB EV).

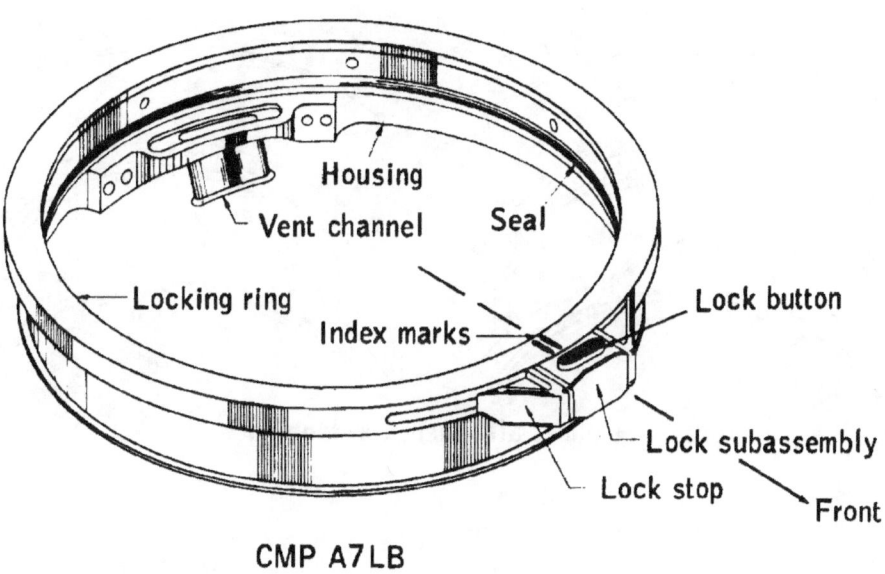

CMP A7LB

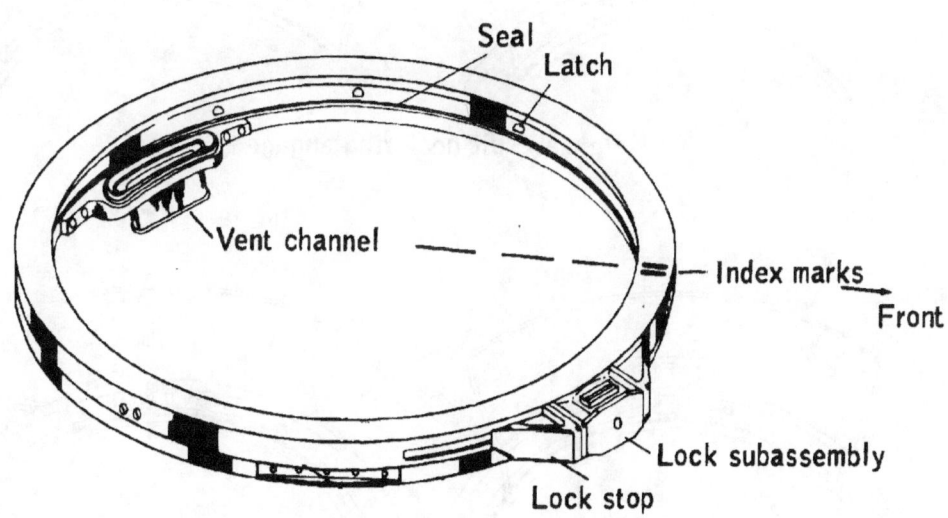

EV A7LB

Figure 2-14.- Helmet attaching neck ring.

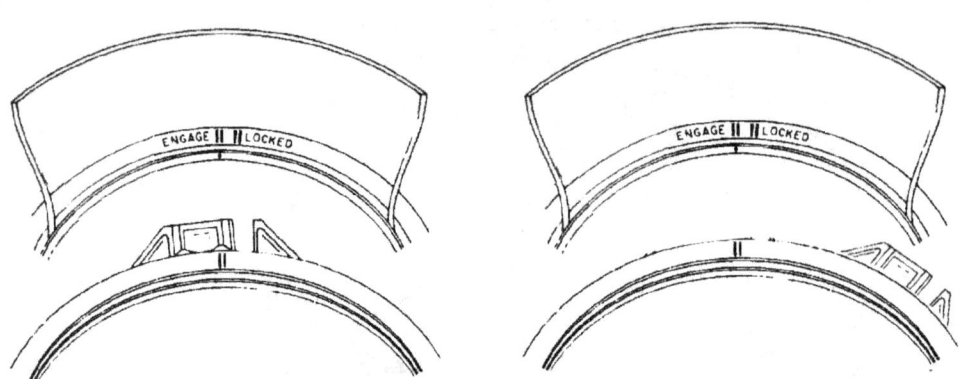
Helmet alinement for donning

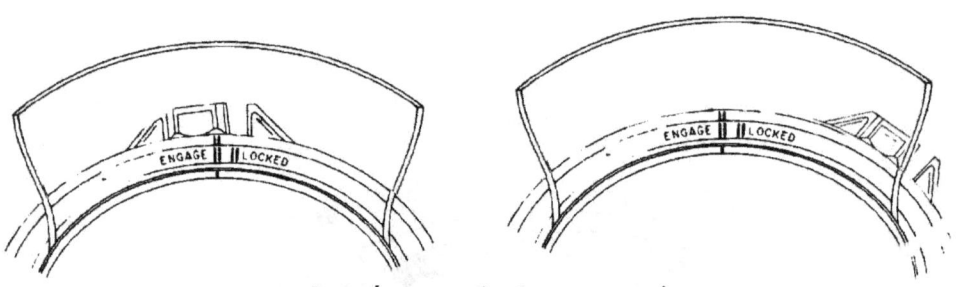
Helmet/suit neck ring engaged

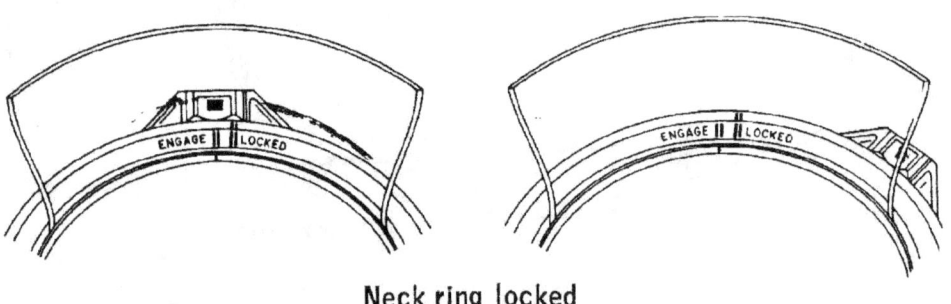

A7LB CMP   Neck ring locked   A7LB EV

Figure 2-14.- Concluded.

printed labels on the helmet neckring identify the ENGAGE and LOCKED positions and facilitate alinement and engagement with the TLSA neckring. Positive locking of the helmet-to-TLSA coupling is ensured by a TLSA-mounted locking ring which is rotated by hand to the engaged, locked, or release positions. A pushbutton lock on the TLSA locking ring permits rotation of the locking ring to the LOCKED position and prevents accidental unlocking. The helmet is donned with the TLSA locking ring in the ENGAGE position by alining and pressing the helmet into place until the latches catch. The helmet is then locked into place by pressing the pushbutton on the TLSA locking ring, sliding the pushbutton lock outward, and rotating the TLSA locking ring to the LOCKED position. The helmet is removed by pressing the pushbutton on the TLSA locking ring, sliding the pushbutton lock outward, and rotating the TLSA locking ring past the ENGAGE position to the release position. When the TLSA locking ring is released at the helmet release position, it returns automatically to the ENGAGE position.

2.3.3.4   Wrist Disconnects

The PGA wrist disconnect (fig. 2-15) coupling includes a suit (female) half and a glove (male) half. The female coupling incorporates a manually actuated lock and unlock mechanism, which has three positions, ENGAGE, LOCK, and UNLOCK. The male half incorporates a sealed bearing which permits 360° glove rotation. The male half of the disconnect is engaged to the female half by alining the glove-half coupling and placing it into the suit-half coupling with the locking ring in the ENGAGE position, then rotating the locking ring to the LOCK position. The glove-half coupling is disengaged or removed from the suit-half coupling by depressing the locklock button with the index finger, and with the thumb and second finger, pulling the two locking tabs from the LOCK position and rotating the locking ring to the open (UNLOCK) position.

2.3.3.5   Gas Connectors

Two inlet and two outlet gas connectors (fig. 2-16) permit the exchange of vent system umbilicals without interrupting the flow of gases to and from the suit. All inlet gas connectors and mating umbilical connectors are anodized blue, and all outlet connectors and mating umbilical connectors are anodized red to preclude reversed connections.

Amendment 2
11/5/71

2-44 CSD-A-789-(1) REV V

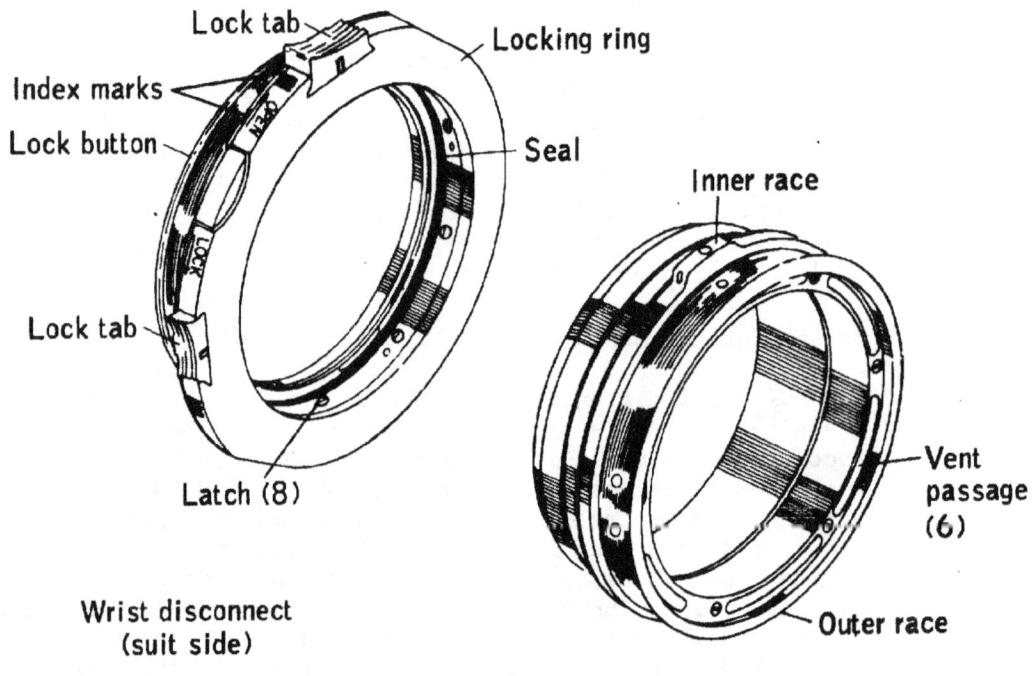

Figure 2-15.- Wrist disconnects.

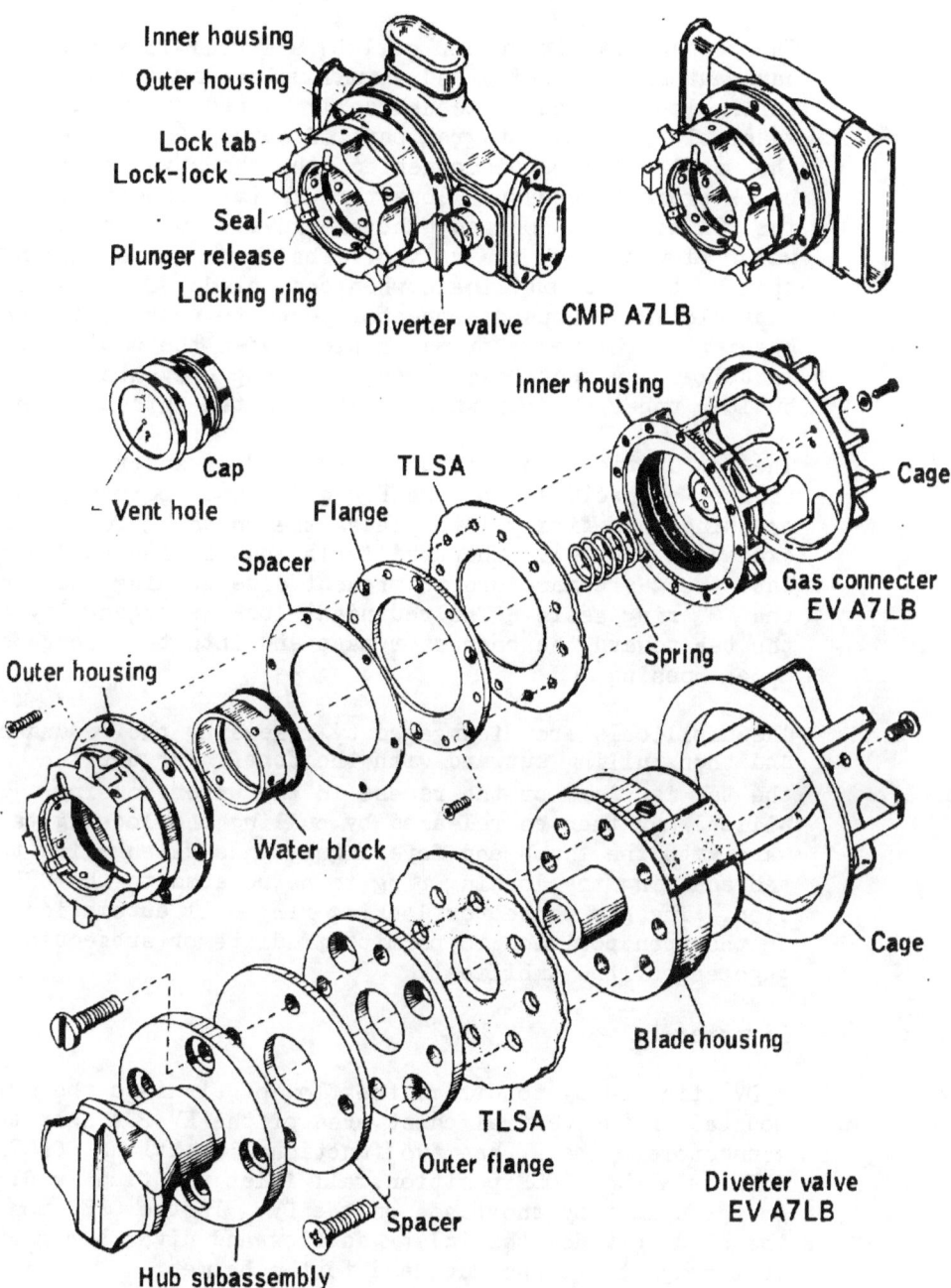

Figure 2-16.- Gas connectors and diverter valve.

The connectors (inlet and outlet) are ball-lock devices and have automatic locking and manual unlocking features. A spring-loaded-closed, mechanically-opened check valve or water block is an integral part of each gas connector. When the umbilicals are disconnected, the check valve or water block prevents pressure loss through the connector. Should the PGA become submerged, the check valve will also prevent water flow through the valves. The check valves are held open by the gas umbilicals when connected. Gas connector caps block the unused connector ports to prevent inadvertent opening of the valve or water block when the umbilicals are not installed. A vent hole through the cap prevents a pressure buildup under the cap when it is inserted into the connector.

The ventilation umbilicals are engaged by inserting the umbilical connectors into the PGA gas connector openings and pressing them firmly into place (the engaging force does not exceed 20 pounds). The umbilicals must be inserted straight into the gas connectors to prevent side loading and damage to the "O" ring seals. The redundant lock is engaged by sliding the tab toward the connector base and into the recess of the upper housing.

The umbilicals are disengaged by releasing the redundant lock and then pulling outward with the forefinger until the tab is clear of the recess in the upper housing. The umbilical may then be released by pulling the locking tabs outward with the thumb and forefinger, thus disengaging them and enabling the locking ring to be rotated to the OPEN position. The gas connector locking ring will automatically lock in the open position to permit immediate or subsequent reengagement of the umbilical.

2.3.3.6   Diverter Valve

A DV (fig. 2-16) to direct the flow of air into the suit is mounted in the central chest area of the EV PGA near the gas connectors. The DV has two functional positions, CLOSE and OPEN. In the CLOSE position, all inlet gas flow is directed to the helmet by the blade on the DV. In the OPEN position, the blade divides the inlet gas flow and diverts a part of it through the torso duct and to the helmet.

A ridged projection on the DV control knob identifies the position of the valve blade. When the ridged projection is vertical (CLOSE position), the blade blocks the passage to the torso duct; when it is horizontal (OPEN position), the blade opens the torso duct passage.

The DV may be rotated 360° in either direction, and spring-loaded, positive (locking) detents are provided at 90° intervals. The valve is operated by pulling out the control knob and rotating it in either direction to the desired position until the locking detent engages.

2.3.3.7 Multiple Water Connector

The multiple water connector (MWC) receptacle (fig. 2-17) includes a double-ball-lock system to engage an LCG dual-passage connector to the inner ball-lock mechanism and a PLSS dual-passage connector to the outer ball-lock mechanism. A plug inserted into the receptacle and locked in place replaces the LCG connector when the LCG is not worn. The plug extends through the receptacle to aline it with the outer surface of the suit.

The inner mechanism is a manually actuated locking and unlocking device. With the locking ring in the OPEN position, the LCG connector is alined with the receptacle port, positioned with the thumb and forefinger, and rotated to the LOCKED position.

The LCG connector is disengaged by pulling out the two locking tabs with the thumb and forefinger and rotating the locking ring to the OPEN position. The LCG connector may then be extracted from the receptacle.

To engage the PLSS connector, the connector must be alined with the port of the receptacle and placed into the receptacle (engaging force should not exceed 20 pounds). The locking mechanism will automatically lock the connector in place. The connector position may be engaged in 180° increments to facilitate convenient connection in the LM.

The PLSS connector may be disengaged by pulling the two locking tabs out and rotating the locking ring to the OPEN position. The locking mechanism will then remain in the OPEN position, ready for immediate or subsequent reengagement.

2.3.3.8 Urine Transfer Connector

The urine transfer connector assembly (fig. 2-18) consists of a PGA-mounted, ball-lock connector and a sized length of interconnecting hose. The connector is flange mounted to the right-leg thigh cone of the PGA where it mates with the urine transfer umbilical of the spacecraft management system. The hose assembly is mounted to the connector on the inside of

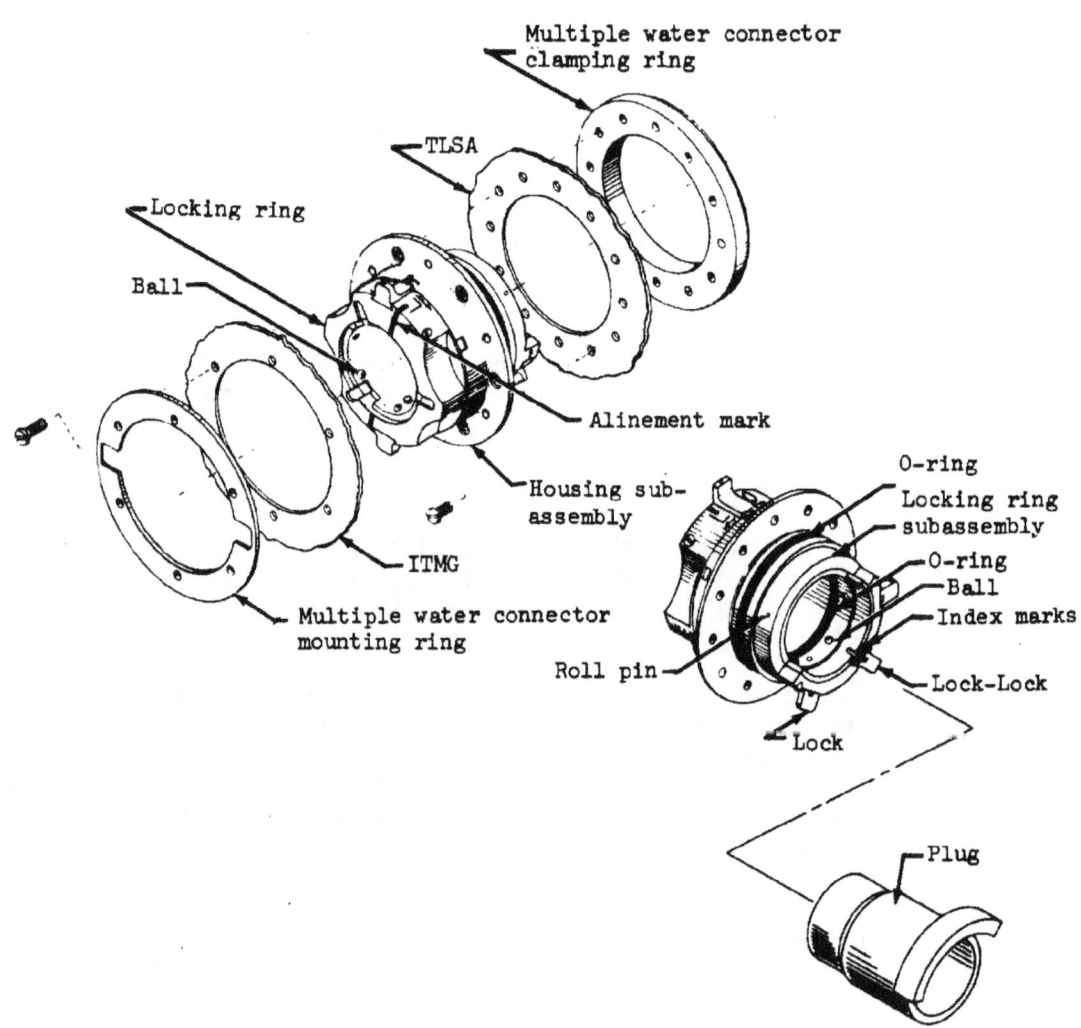

Figure 2-17.- Multiple water connector.

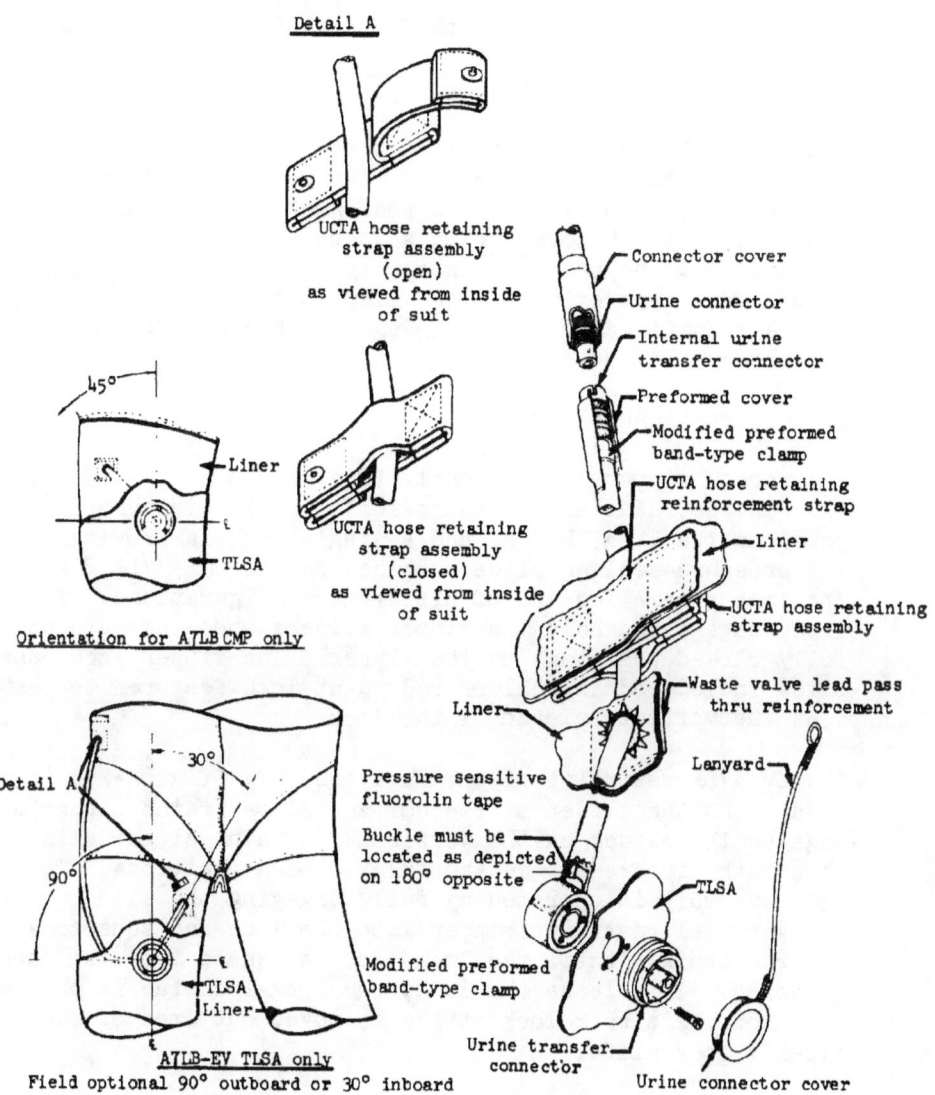

Figure 2-18.- Urine transfer connector.

the PGA, and it extends to a male adapter which mates with
the UCTA connector. The assembly transfers urine from the
UCTA to the spacecraft waste management system. A preformed
rubber connector cover is fitted over the mated UCTA/TLSA
urine transfer hose connector to improve comfort and to preclude possible abrasion to the TLSA bladder.

2.3.3.9   Biomedical Injection Patch

A circular biomedical injection patch (fig. 2-19) is sewn to
the left-thigh cone of the PGA. The patch is made from a
silicone rubber disk which is self-sealing to permit a crewman to inject a hypodermic in a vacuum environment without
jeopardizing the pressure integrity of the PGA. The patch
is placed at approximately the midpoint of the PGA thigh cone
and is identified by a red zigzag stitch line around the
perimeter.

2.3.3.10   Zipper Lock Assemblies

A separate zipper lock assembly (fig. 2-20) is provided for
the PGA restraint and pressure-sealing slide fasteners (zippers) on the EV A7LB PGA, and a single lock is provided for
the pressure-sealing slide fastener on the CMP A7LB PGA.
The lock assemblies are of different configurations. The
locks engage and hold the zipper sliders when they are at the
fully closed positions on the zipper. The zipper lock assemblies include additional or redundant lock features to prevent inadvertent release of the lock.

The EV A7LB restraint zipper lock assembly (fig. 2-20) is
mounted on the slider of the horizontal restraint zipper and
engages the slider on the vertical restraint zipper slider
when both zippers are in the fully closed positions. The
lock assembly is operated by fully engaging the slider of
the vertical restraint zipper into the lock and squeezing
the red striker until the lock-lock tab snaps into the lock
position. To release the lock, the lock-lock tab is pulled
out, and the zipper lock strike is moved out free of the vertical zipper slide.

The EV A7LB pressure zipper lock assembly (fig. 2-20) for
the pressure sealing zipper is mounted on the CLOSE zipper
stop. When the zipper is fully closed, the slider depresses
the safety plunger which permits the lock to be actuated.
The lock is actuated by pressing inward on the safety shaft
while simultaneously turning it until the spring retaining
pin is moved fully into the detent slot. To disengage the

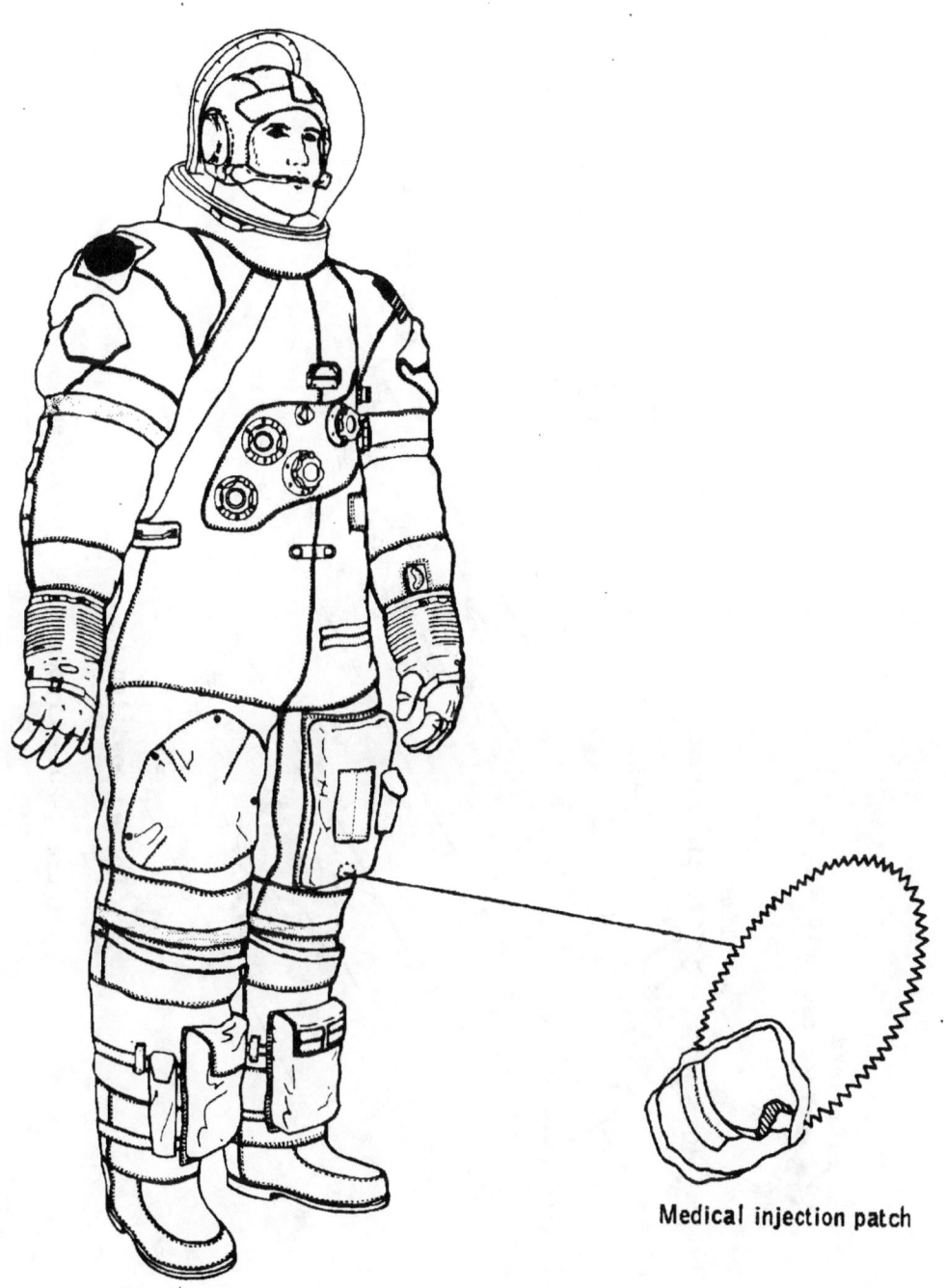

Figure 2-19.- Medical injection patch.

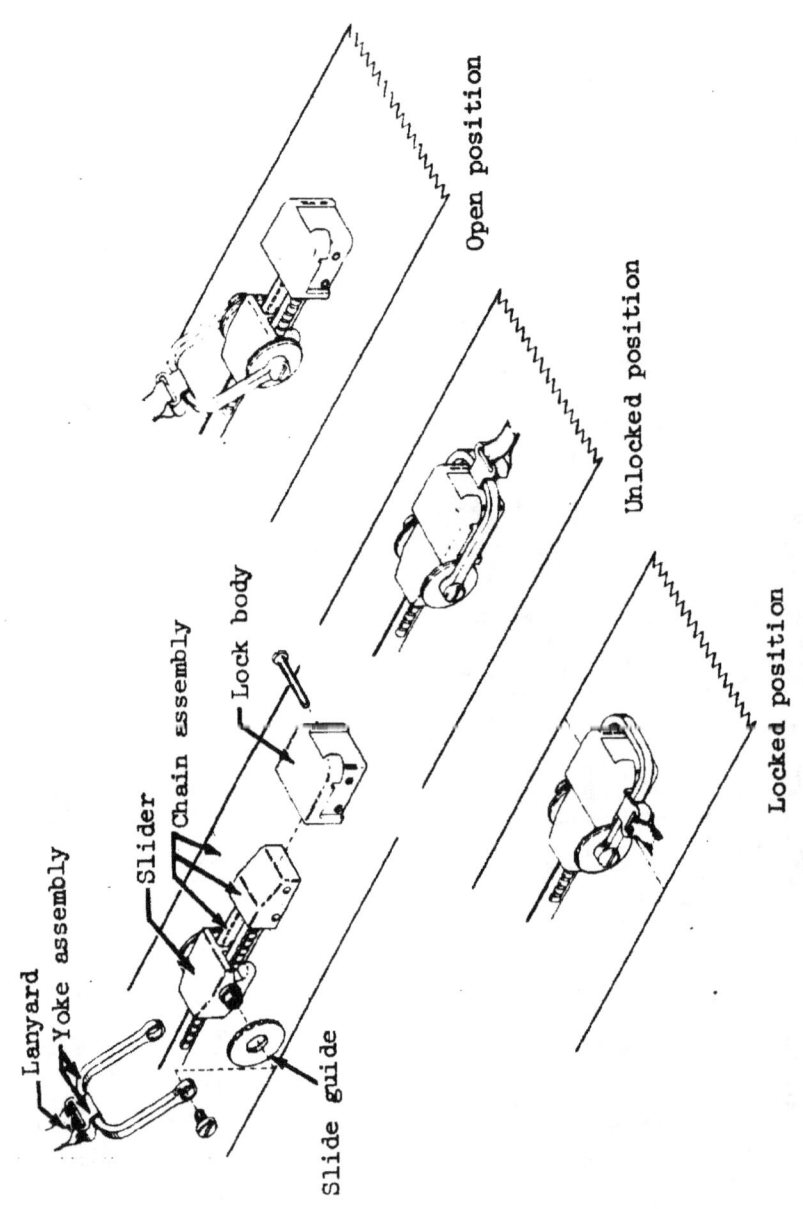

Figure 2-20.- Zipper lock assemblies.

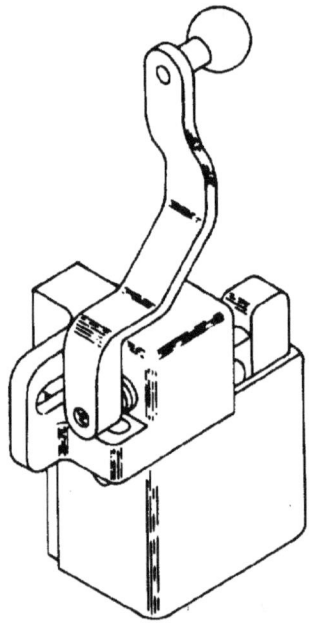
EV A7LB pressure zipper lock assembly

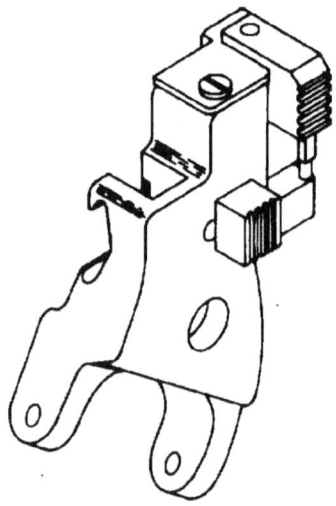
EV A7LB restraint zipper lock assembly

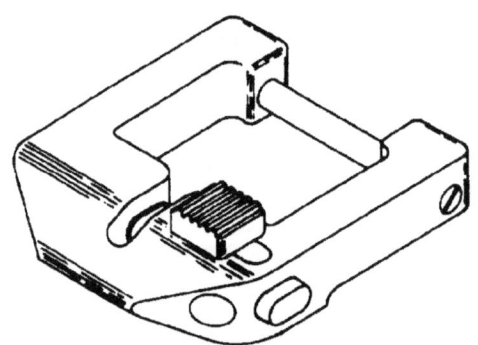
CMP A7LB slide fastener lock assembly

Figure 2-20.- Zipper lock assemblies.

lock, the locking shaft is depressed and the safety arm is rotated out and away from the zipper allowing the locking shaft to disengage the zipper strike.

The CMP A7LB pressure-sealing slide fastener lock assembly (fig. 2-20) holds the slider of the pressure-sealing closure to prevent accidental opening. The lock assembly may be placed in two positions, LOCK and UNLOCK. The LOCK position is achieved by pushing the lock slider inboard to the stop using the thumb and forefinger. The red slider should not protrude beyond the body of the assembly when the slider is in the LOCK position. An OPEN position is achieved by pushing the lock assembly release button outboard of the stop using just the thumb. To engage the lock to the pressure-sealing closure slider, the lock assembly is firmly pulled over the slider and then the assembly is locked. The slide fastener closure is released by unlocking the lock assembly and lifting the lock assembly away from the pressure-sealing closure slider. A detent assembly holds the lock assembly slider in the LOCK and UNLOCK positions.

2.3.3.11 Pressure Relief Valve

The pressure relief valve (fig. 2-21) relieves suit pressures in excess of 5.0 psid. Relief cracking limits are 5.0 to 5.75 psid. The valve will reseat as suit pressure reduces to 4.6 psid and shall not leak more than 4.0 scc per minute when closed at 4.6 psid. The valve accommodates a relief flow of 12.2 lb/hr minimum at 5.85 psia in the event of a faulted-open primary oxygen pressure regulation in the PLSS.

The pressure relief valve may be blocked to preclude the relief of suit pressure or to stop leakage through the valve. A cap fitted over the valve and locked in place by a cam lock system blocks the exhaust ports to prevent pressure relief through the valve.

Amendment 2
11/5/71

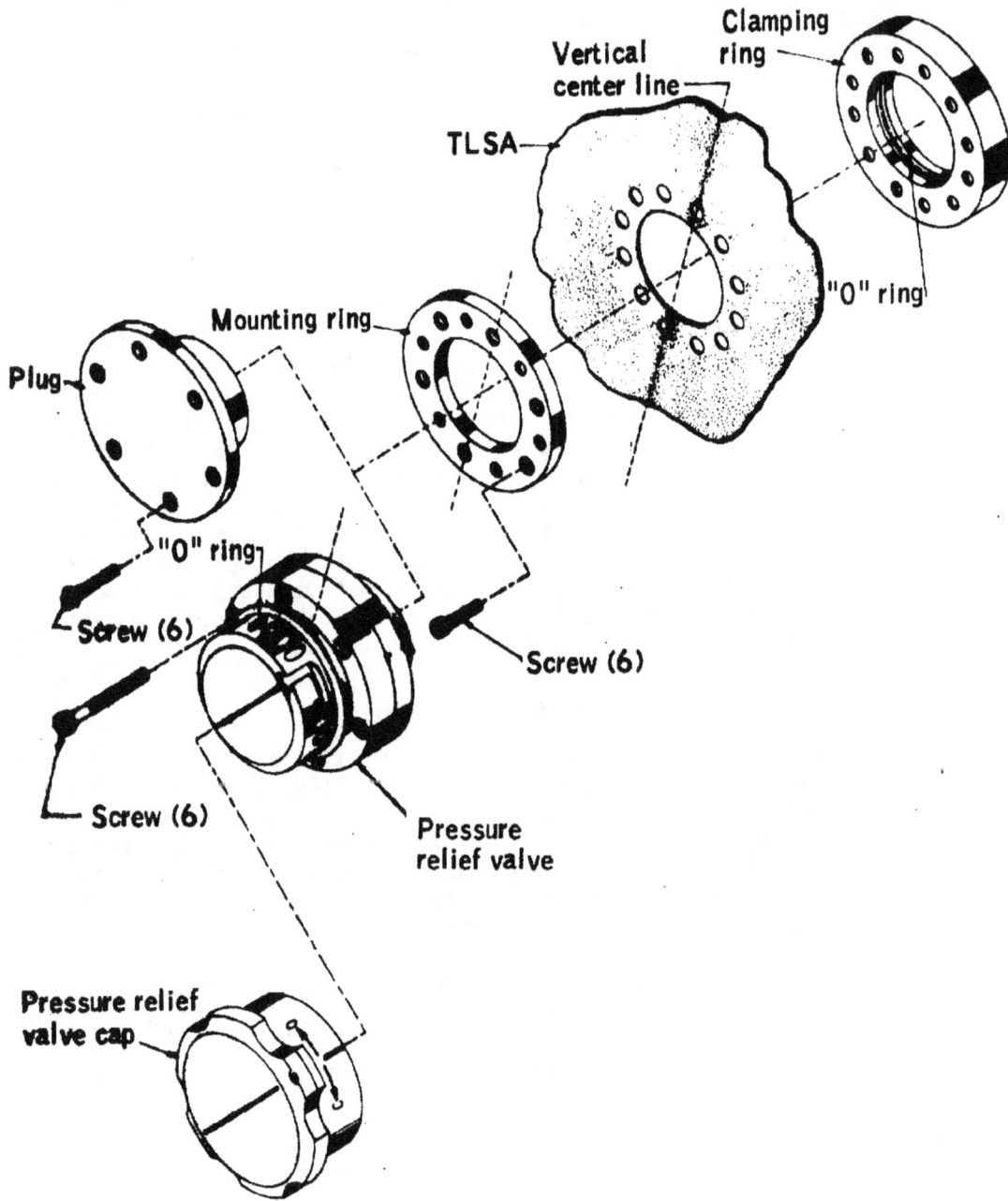

Figure 2-21.- Pressure relief valve.

2.3.3.12   Biomedical Belt

The biomedical belt (fig. 2-22) supports the signal conditioners and power converter as a part of the biomedical instrumentation system. The power converter is located in the right-hand pocket (as worn), the ECG signal conditioner in the center pocket, and the impedance pneumogram (ZPN) signal conditioner in the left-hand pocket. The connector ends of these units are colored red, blue, and yellow, respectively. When installing or reinstalling the units, the above order is maintained to assure that proper signal path connections are made. When the belt is transferred between the LCG and CWG, the color-coded electrode harnesses are disconnected at the units, and the units are retained in the belt. The biomedical harness need not be disconnected from the belt. The electrodes are not removed to change garments.

2.3.3.13   Biomedical Harness

The biomedical harness (fig. 2-22) is a four-branch assembly that interfaces with the two biomedical instrumentation signal conditioners (ECG and ZPN), the dc-dc power converter, and the main branch which mates with the suit electrical harness. The wires are covered with a sheath of Teflon fabric anchored to each connector by nylon wrapping cord. The harness is held in place by the biomedical belt and, through its mechanical connectors, with the dc-dc converter and the signal conditioner.

2.3.3.14   Suit Electrical Harness

The suit electrical harness (fig. 2-22) has a central 61-pin connector from which two branches extend. One branch connects to the communications cap or carrier, while the second, shorter branch connects to the biomedical harness. The communications branch has a 21-pin connector, and the biomedical instrumentation branch has a 9-pin connector. A groove machined into the mounting face of the central 61-pin connector uses an O ring to provide a seal when the electrical harness is mounted to the TLSA. Each branch is covered with a Teflon fabric sheath. The Teflon fabric sheaths are attached to each connector with wrapping cord and an adhesive. The central 61-pin connector receives the ball/lock engagement mechanism of the communications and biomedical instrumentation umbilical of the spacecraft or the PLSS. The 9- and 21-pin connectors employ a dual-pawl or latch-engaging mechanism.

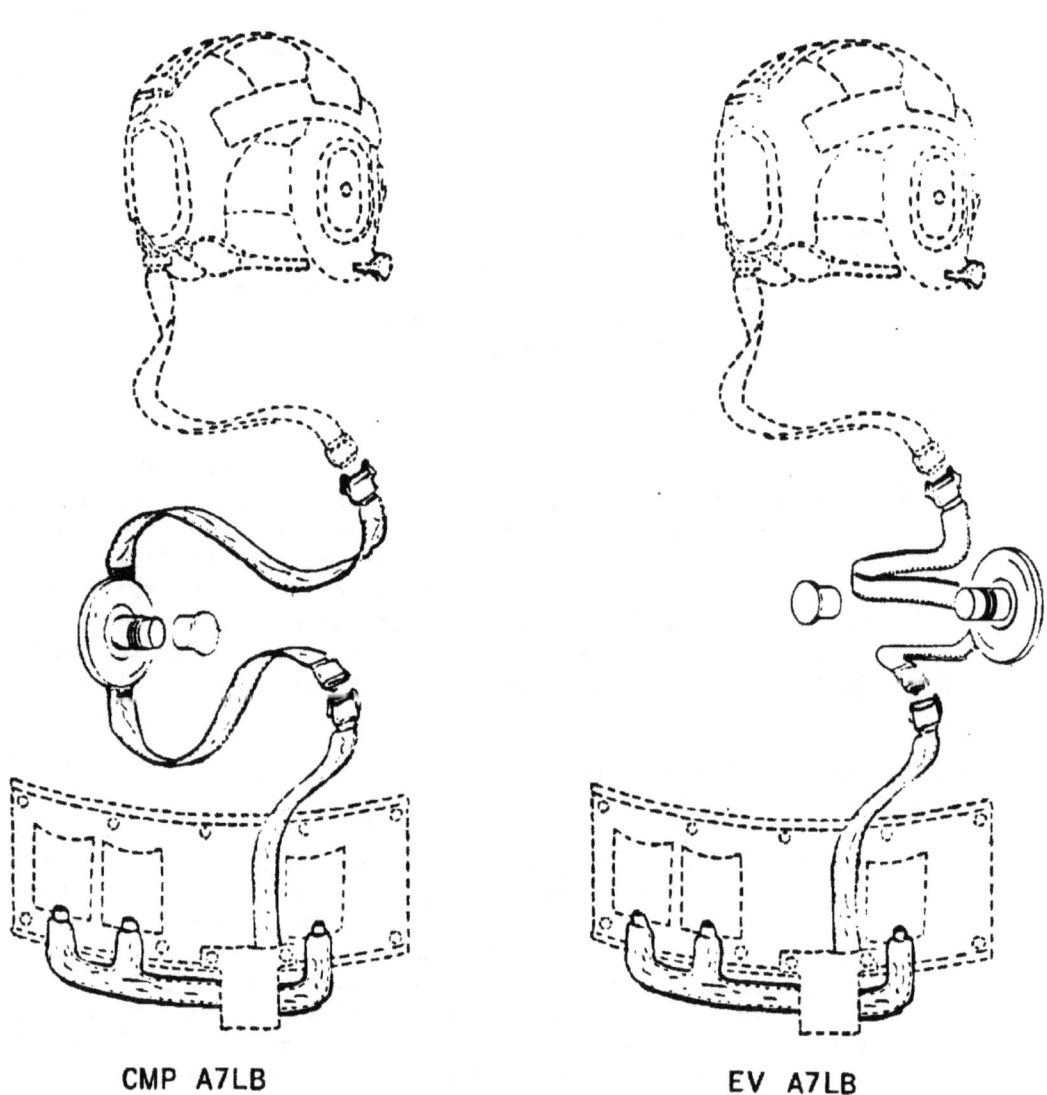

Figure 2-22.— Biomedical and suit electrical harness and biomedical belt.

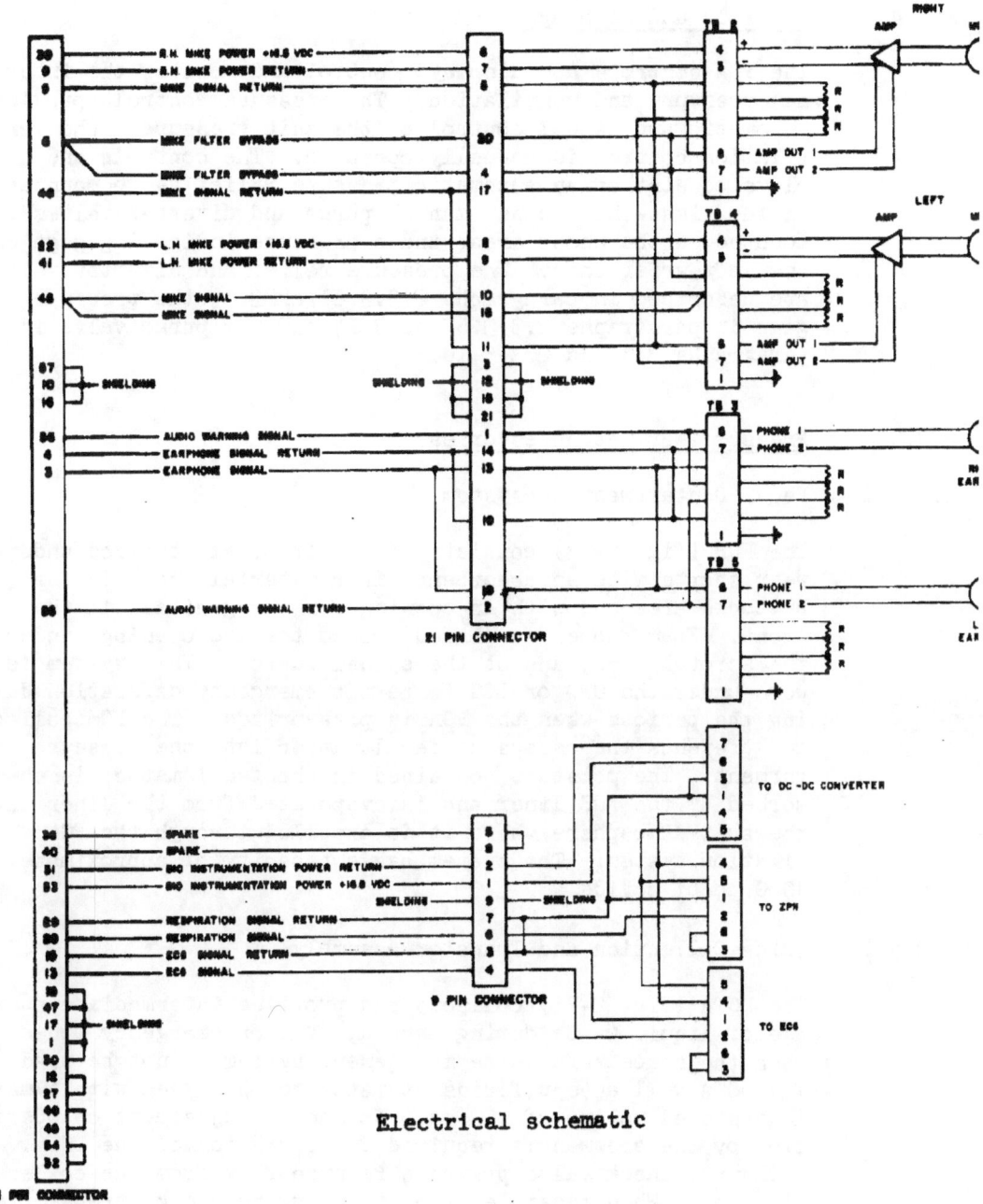

Electrical schematic

Figure 2-22.- Concluded.

2.3.4 Controls and Displays

The PGA controls and displays control and monitor the internal pressure and ventilation. The pressure controls provide automatic and manual control of the suit pressure. The ventilation control is manually operated. The controls and displays consist of an automatic pressure relief valve mounted on the right-thigh cone, manual purge and diverter valves mounted on the chest area, and a pressure indicating gage on the left-wrist cone. The pressure relief and diverter valves are described in paragraphs 2.3.3.11 or 3.1, the pressure gage in paragraphs 2.3.3.6 and 3.1, and the purge valve in paragraphs 3.1 and 2.3.5.10.

2.3.5 Pressure Garment Accessories

2.3.5.1 Fecal Containment Subsystem

The FCS (fig. 2-23) consists of a pair of elasticized underwear shorts with an absorbent liner material added in the buttocks area and with an opening for the genitals in the front. Foam rubber is placed around the leg opening, under the scrotal area, and at the spinal furrow. This system is worn under the CWG or LCG to permit emergency defecation during the periods when the PGA is pressurized. The FCS collects and prevents the escape of fecal matter into the pressure garment. The moisture contained in the fecal matter is absorbed by the FCS liner and is evaporated from the liner into the suit atmosphere where it is expelled through the PGA ventilation system. The system has a capacity of approximately 1000 cc of solids.

2.3.5.2 Urine Collection and Transfer Assembly

The UCTA (fig. 2-23) collects and provides intermediate storage of liquid waste during launch, EVA, or emergency modes when the spacecraft waste management system cannot be used. The UCTA will accept fluids at rates to 30 cc/sec with a maximum stored volume of 950 cc. No manual adjustment or operation by the crewman is required for operation of the UCTA. A flapper check valve prevents reverse flow from the collection bag. When feasible, the stored urine can be transferred

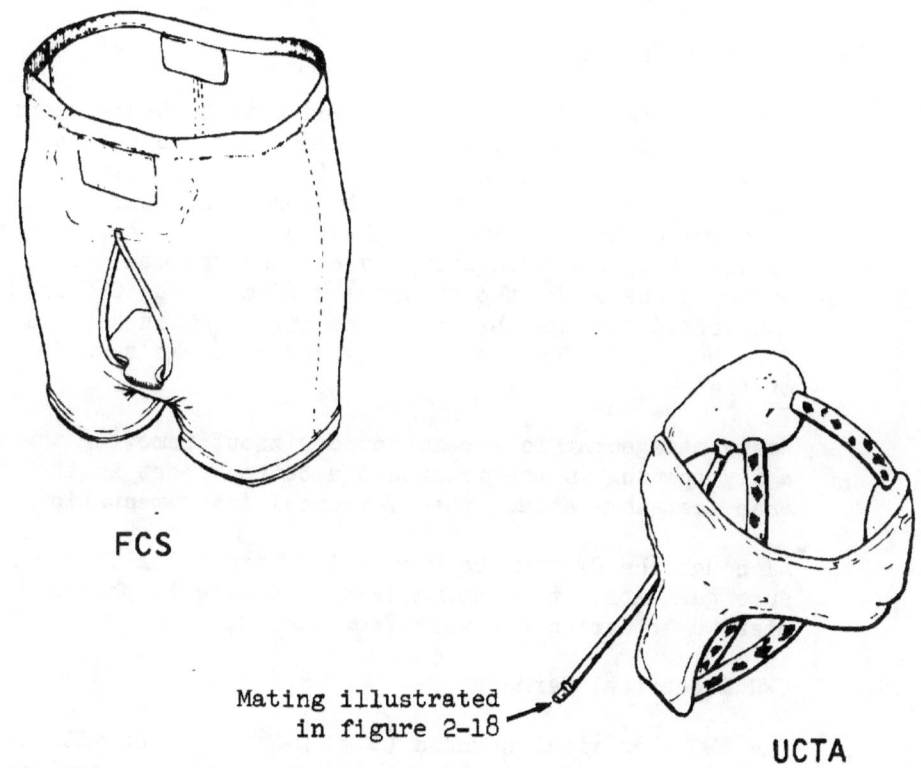

Figure 2-23.- Fecal containment subsystem and urine collection and transfer assembly.

through the suit wall by hose to the CM or LM during pressurized or depressurized cabin operation.

The UCTA is worn over or under the CWG or the LCG and is connected by hose to the urine transfer connector on the PGA. The urine transfer connector is a quick-disconnect fitting used to transfer urine from the UCTA to the spacecraft waste management system. A UCTA transfer adapter is provided on board the CM for use by the crewman to dump the liquid waste after the PGA has been doffed.

2.3.5.3 Constant Wear Garment

The CWG (fig. 2-24) is a one-piece cotton undergarment which is worn next to the skin and encompasses the entire body exclusive of the head and hands. It is worn during IV CM operations for general comfort, to absorb perspiration, and to hold the biomedical instrumentation system. It absorbs excessive body moisture and prevents the crewman's skin from becoming chafed by the pressure garment. The CWG is donned and doffed through the front opening which is kept closed by five buttons. The feet are covered by socks sewn to the legs of the CWG.

Waste management is accommodated without removing the CWG by a fly opening in the front and a buttock port in the rear. Snap fasteners attach the biomedical instrumentation belt.

Although the CWG may be worn under either the CMP or EV pressure garments, it is normally used during IV phases of the mission or during EVA work from the CM.

2.3.5.4 CWG Electrical Harness

The CWG electrical harness (fig. 2-24) is used with the CWG or inflight coverall garment and provides a mechanical and electrical interface with the communications carrier, biomedical harness assembly, and the spacecraft communications umbilical. It replaces the suit electrical harness when the PGA is doffed and the CWG is worn.

The CWG electrical harness consists of a central 61-pin connector from which two branches extend. One branch conducts communications signals while the second, shorter branch connects to the biomedical harness. The communications branch includes a 21-pin connector which interfaces with the communications carrier or lightweight headset. The biomedical instrumentation branch has a 9-pin connector which interfaces

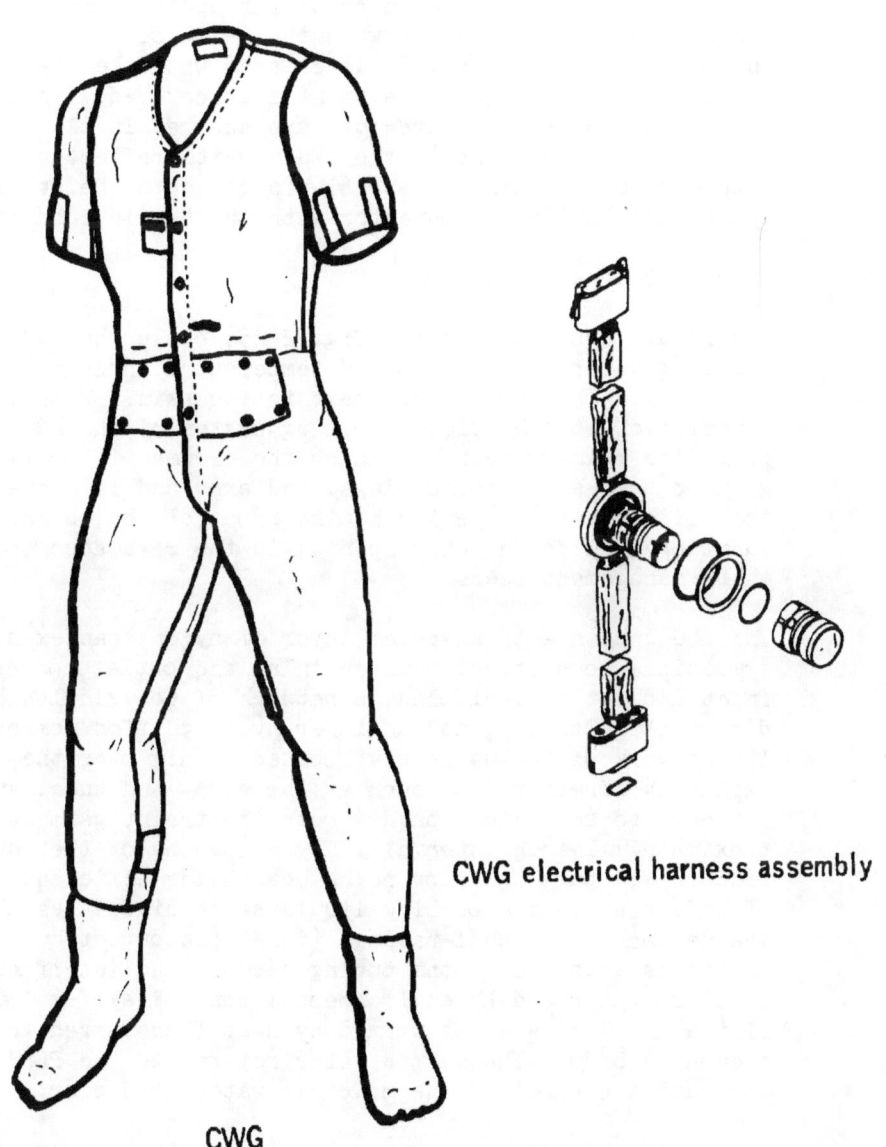

Figure 2-24.- Constant wear garment and electrical harness.

with the biomedical harness. The 61-pin connector protrudes
through the inflight coverall garment at the upper chest area
to engage with the electrical umbilical. An aluminum washer
spacer positions the 61-pin connector housing and ensures
proper depth of engagement when the ball-lock mechanism of
the electrical umbilical is interfaced with the 61-pin connector.
Each branch of the harness is covered with a Teflon
fabric sheath, and the branches are secured in place by two
snap tabs on the front of the CWG. White reflective tape
attached to the shell of the 61-pin connector helps aline the
spacecraft umbilical connector with the 61-pin connector.

2.3.5.5   Liquid Cooling Garment

The liquid cooling garment (fig. 2-25) cools the body by circulating
water at a controlled temperature through a network
of tubing. The LCG is worn next to the skin. When it is
interfaced with the liquid cooling system of the PLSS or LM,
it is the primary means by which the crewman is cooled. The
garment covers the torso, legs, and arms and is donned through
the slide fastener opening in the front of the torso. An
additional slide fastener opening in the rear accommodates
waste management needs.

The LCG consists of an outer layer of nylon spandex material,
a multiple connector for water inlet and outlet connections,
inlet and outlet manifolds, a network of polyvinylchloride
distribution tubing, and an inner nylon chiffon comfort liner.
The network of tubing is distributed evenly over the body,
excluding stress points such as the elbow and knee, and is
stitched to the nylon spandex outer restraint garment at approximately
1-inch intervals. Even spacing of the tubing
network and parallel flow paths permit the efficient transfer
of body heat to the cooling liquid as it circulates through
the network. The dual-passage (inlet and outlet) water connector
is attached to the tubing network and interfaces with
the PLSS water and LM environmental control system (ECS) umbilicals.
The water is warmed by heat transferred from the
crewman's body. The warmed water returns to the PLSS through
the outlet channel of the multiple water connector.

The nylon chiffon liner separates the tubing network from the
body and also contributes to body comfort by absorbing and
evaporating perspiration into the PLSS or ECS oxygen systems.
Comfort pads are installed at strategic points on the LCG.
Custom-sized socks are physically attached to the LCG; however,
the socks do not incorporate cooling tubes. There are
eight snap fasteners located in the abdominal area of the

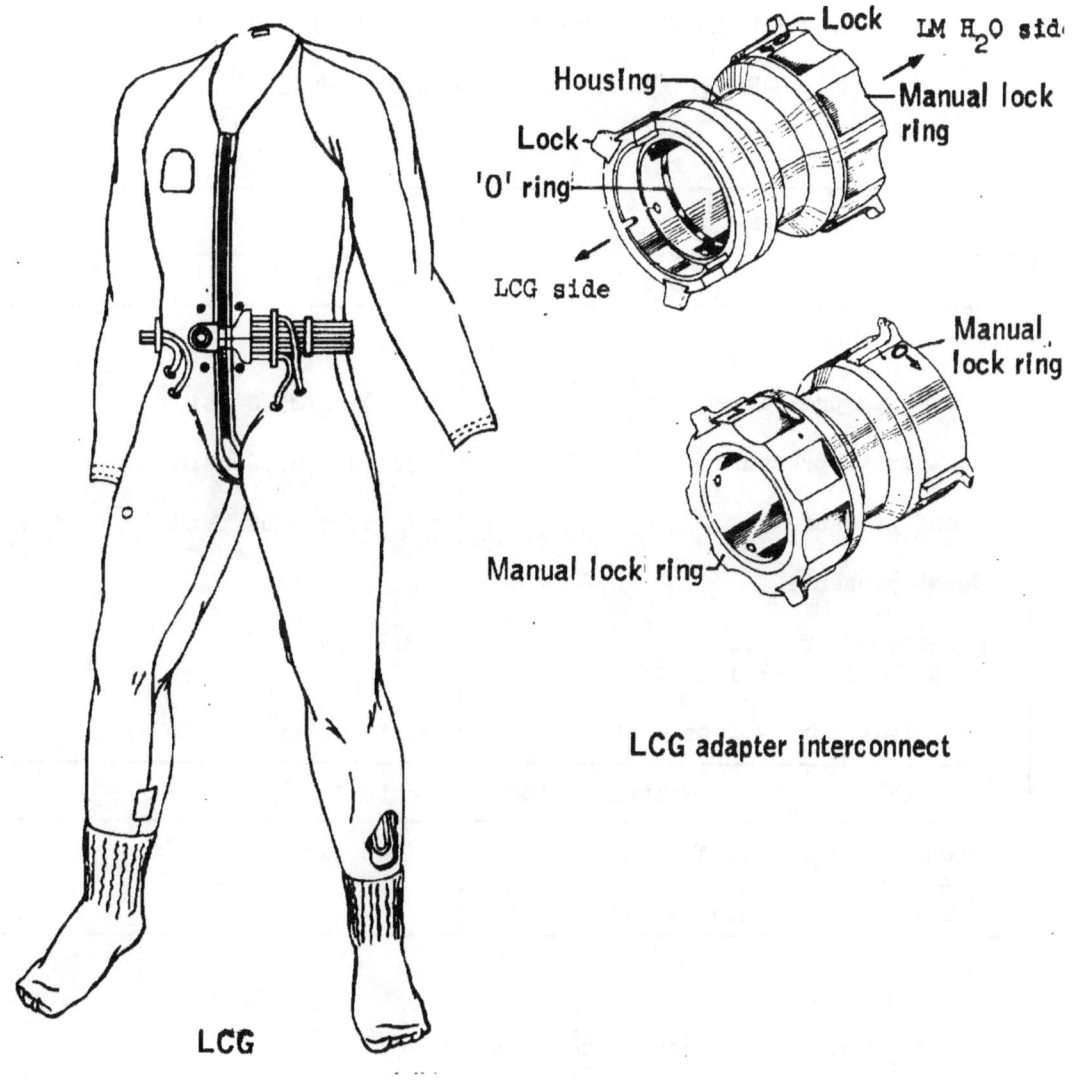

Figure 2-25.- Liquid cooling garment and LCG adapter interconnect.

garment to secure the biomedical belt. Three passive dosimeter pockets are placed at strategic points about the garment.

Table 2-IX lists the main characteristics of the LCG and the multiple water connector.

TABLE 2-IX.- PERFORMANCE CHARACTERISTICS OF THE LIQUID COOLING GARMENT AND MULTIPLE WATER CONNECTOR

Item	Value
Liquid cooling garment	
Weight (charged)	7.00 lb[a]
Operating pressure	4.20 to 23.0 psid
Structural pressure	31.50 ± 0.50 psid
Proof pressure	31.50 ± 0.50 psid
Burst pressure	47.50 psid
Pressure drop (4.0 lb/min at 70° ± 10° F inlet)	3.35 psi[b]
Leak rate for 19.0 psid at 45° F	0.58 cc/hr
Multiple water connector	
Pressure drop (4.0 lb/min at 70° ± 10° F, both halves, both directions)	1.45 psi

[a] Design value.
[b] Includes both halves of connector.

2.3.5.6   LCG Adapter Interconnect

The LCG connector adapter interconnect (fig. 2-25) is a dual-ball lock adapter which permits an interface between the LCG and LM liquid cooling systems when the PGA is removed. The assembly employs manual locking and unlocking mechanisms for engaging and disengaging both liquid cooling system connectors. The inflight coverall garment is normally worn over the LCG during IV activity, supports the LCG LM umbilical, and precludes kinks and water restrictions in the tubing.

2.3.5.7 Insuit Drinking Device (ISDD)

The insuit drinking device (fig. 2-26) provides approximately 32 ounces of potable water within the PGA during lunar surface extravehicular activities. The ISDD consists of a flexible film bag with an inlet valve for filling and an outlet tube and tilt valve for drinking. The bag is attached between the PGA bladder and liner at the neck ring by means of hook and pile Velcro. The bag is filled with potable water from the spacecraft water system by means of the water dispenser/fire extinguisher.

Amendment 2
11/5/71

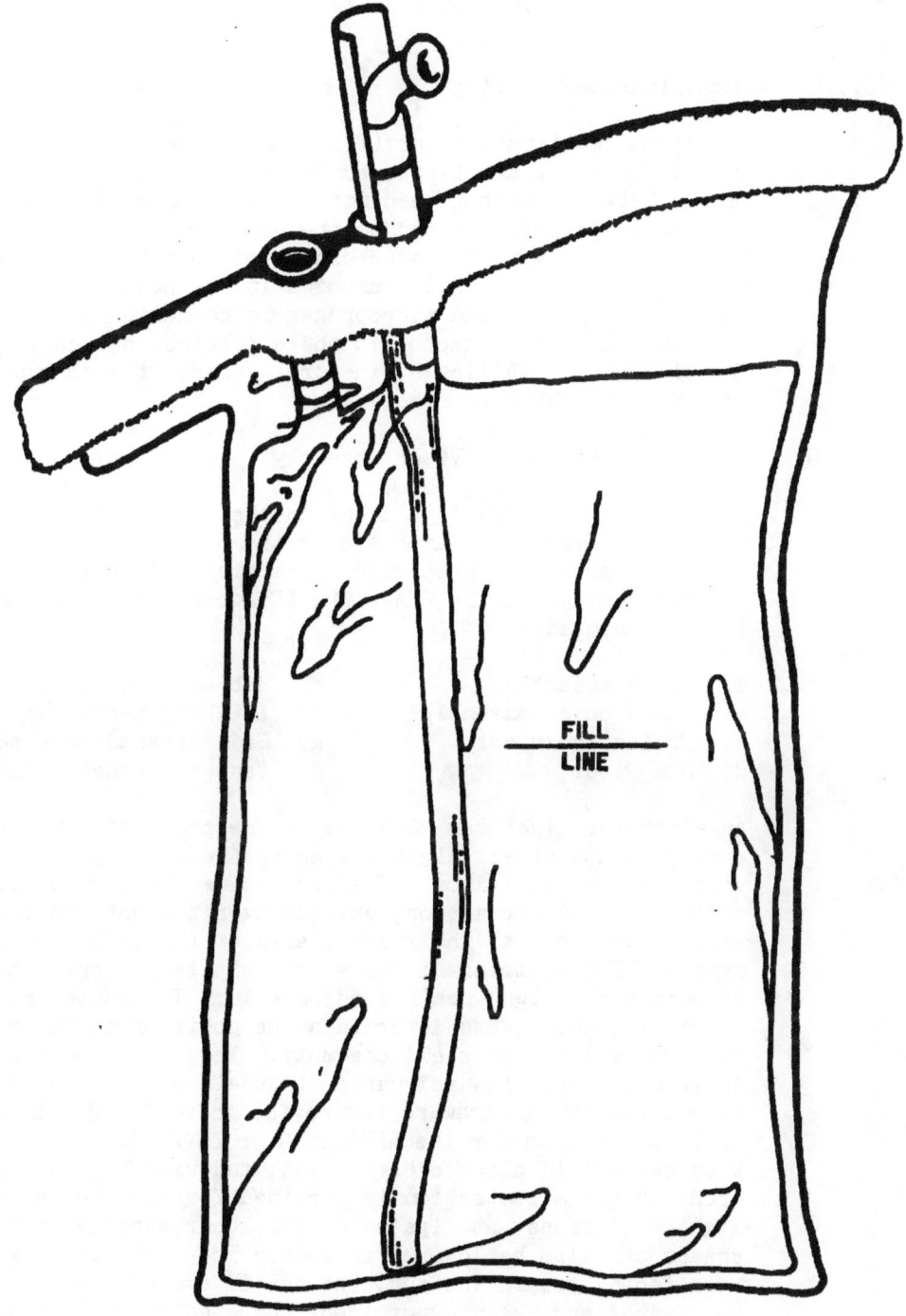

Figure 2-26.- Insuit drinking device.

Amendment
11/5/71

2.3.5.8 Communications Carrier

The communications carrier (fig. 2-27) provides microphones and earphones in a soft-suspension skull cap. Acoustic isolation between earphone and microphone is achieved when the carrier is properly fitted to the wearer. The connection may be made directly to the spacecraft communications system or through the PGA internal communication harness. The wiring from the earphones and microphones is connected by a flat pigtail to a 21-pin connector in the electrical harness assembly. The electrical umbilicals, in turn, connect the communications system to the PLSS or spacecraft.

2.3.5.9 Lunar Extravehicular Visor Assembly

The LEVA (fig. 2-28) is a light and heat attenuating assembly which fits over the clamps around the base of the PHA. It provides additional protection from micrometeoroids and accidental damage to the PHA. The LEVA consists of the following subassemblies.

a. Shell assembly
b. Shell cover assembly
c. Protective visor
d. Sun visor
e. Hub assemblies (2)
f. Latching mechanism
g. Side eyeshade assemblies (2)
h. Center eyeshade assembly

An elastomer light seal located on the protective visor stiffener prevents direct light leakage between the protective visor and the sun visor. The protective visor, when lowered to the full-DOWN position, extends over a light and thermal seal arrangement at the frontal area of the shell cover assembly. The position of the visors within the shell assembly and about the light seal is adjustable. The radial position of visor support cams determines the position of the visors with respect to the shell assembly. The shell cover assembly is attached over the polycarbonate shell and extends below the helmet attaching hardware to provide thermal and micrometeoroid protection for the LEVA/ITMG or LEVA/CLA interface area. When secured in place over the PHA, and with both visors lowered, adequate protection is provided for the thermal and light conditions anticipated on the lunar surface. The eyeshades can also be lowered to reduce low-angle solar glare. When facing toward the sun, the center eyeshade assembly may be lowered and the viewport door adjusted to provide additional solar glare protection.

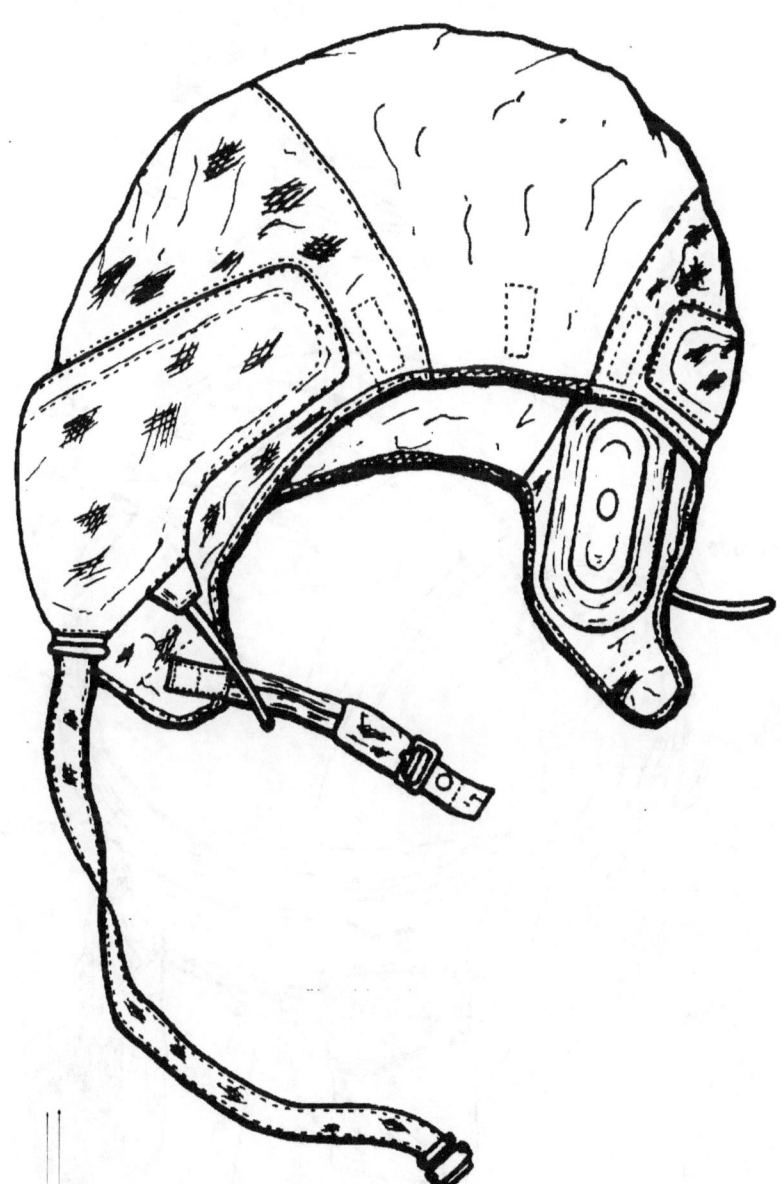

Figure 2-27.- Communications carrier.

2-70 CSD-A-789-(1) REV V

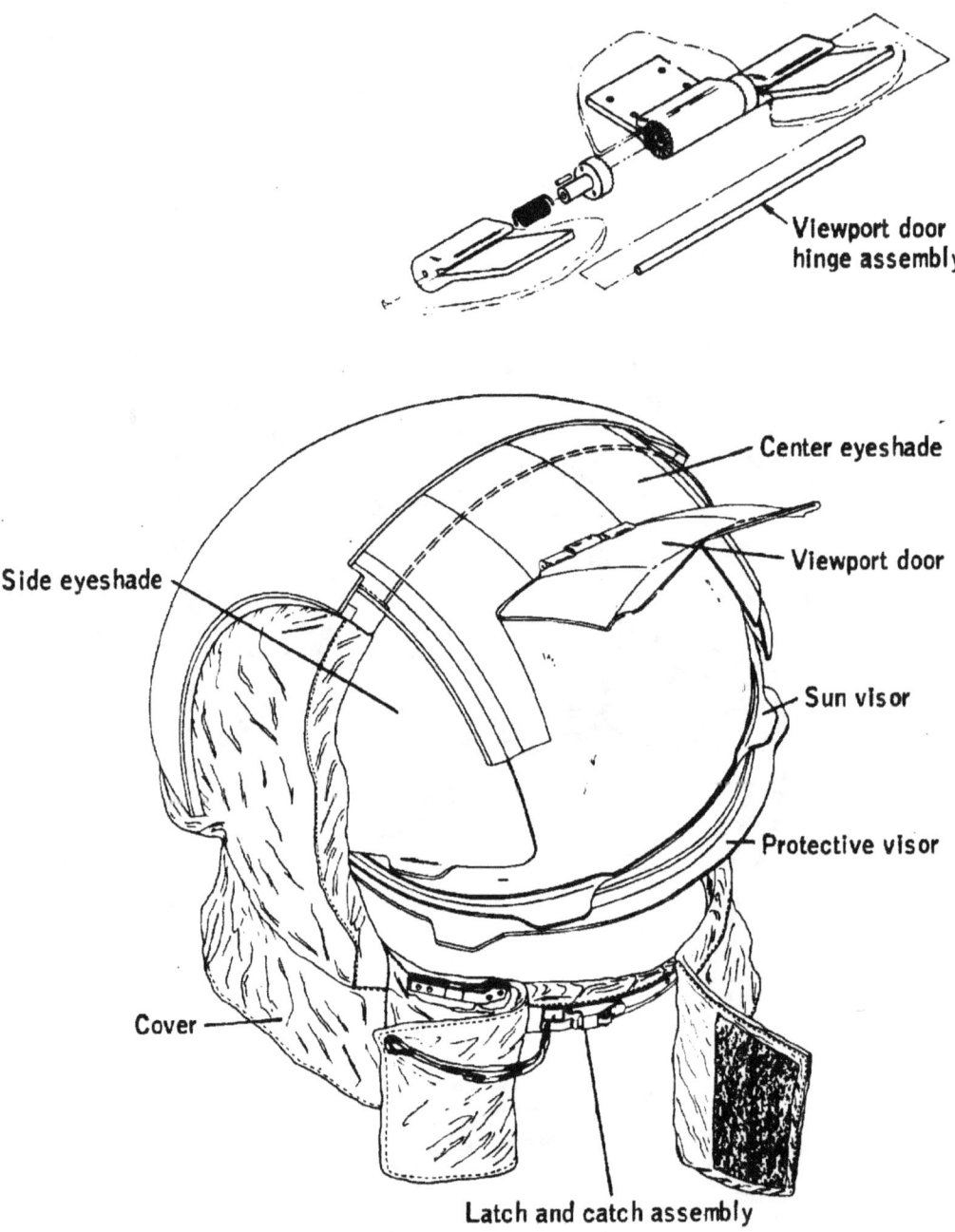

Figure 2-28.- Lunar extravehicular visor assembly.

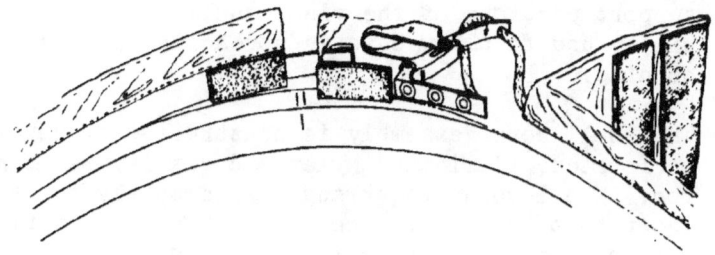

LEVA to neck ring before latching

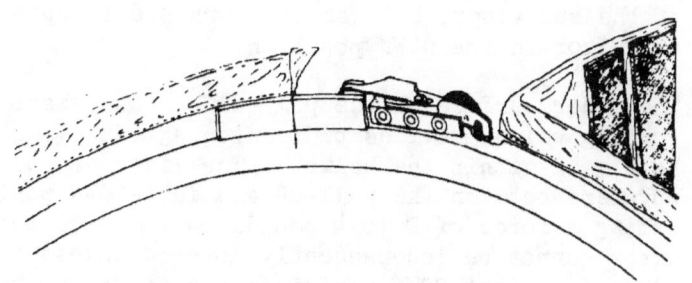

LEVA to neck ring after latching

Figure 2-28.- Concluded.

The shell assembly is a formed polycarbonate structure to which the visors, hinge assemblies, eyeshades, latch, and shell cover assembly are attached. The shell assembly latches around the pressure helmet at the neck ring, and, when the latch is secured, a rigid connection between the two assemblies is assured. Adjacent to the visor hinge, straps constructed of polypropylene are employed across the cut-out support portions of the visor shell to permit flexual durability and to allow ease in spreading the visor during LEVA donning.

The shell cover assembly is constructed of seven layers of perforated, aluminized Mylar and six layers of nonwoven Dacron. The layers are arranged alternately to reduce interlayer heat transfer. The outer layer or covering is made of Teflon-coated Beta yarn for additional thermal and fire protection. Potential scuff areas on the forward edge are reinforced with Teflon fabric. Flameproof hook-and-pile fastener tape (Velcro) is used to attach the collar over the LEVA/ITMG or LEVA/CLA interface area.

The protective visor is an ultraviolet-stabilized polycarbonate shield which affords impact, micrometeoroid, and ultraviolet ray protection. It can be positioned anywhere between the full-UP and full-DOWN positions and requires a force of 2 to 4 pounds for movement. A coating is added to the inner surface of this assembly. The elastomer seal on the upper surface of the stiffener prevents light passage between the two visors. The protective visor can be lowered independently of the sun visor, but cannot be raised independently with the sun visor in the DOWN position.

The inner surface of the polysulfone sun visor has a gold coating which provides protection against light and reduces heat gain within the helmet. The visor can be positioned anywhere between the full-UP and full-DOWN positions by exerting a force of 2 to 4 pounds on the pull tabs. The sun visor cannot be independently lowered unless the protective visor is in the DOWN position, but it can be raised or lowered independently when the center eyeshade is in the full-UP position and the protective visor is in the DOWN position.

The hinge assemblies located on each side of the LEVA shell are support and pivot devices for the two visors and eyeshades. The hinge positions adjust for a proper fit of the visors to the shell and helmet assemblies and to aid in achieving a good light seal. Each hinge assembly is comprised of a bolt extending through a two-piece hub arrangement which supports dissimilar-material washers, the spacers, and a spring. Tension on the spring is adjustable and determines the force necessary for visor and side eyeshade movements. After adjustment, the hinge bolt is safe-tied with lock wire.

The latching mechanism is constructed of stainless steel and is used to secure the base of the LEVA shell around the PHA above the helmet neck ring. The over-center feature of the latch pulls the two sides of the front portion of the LEVA shell structure together and tightens it around the PHA. A lanyard attached to the actuating tab of the latch and the shell cover assembly permits easy actuation of the latch with a gloved hand. The lanyard is visible when the collar is held open.

The eyeshade assemblies are constructed of fiberglass and are coated with white epoxy paint on the outer surfaces. The inner surfaces are coated with black epoxy paint. The side eyeshades are attached to the hinge assemblies and can be lowered independently of the sun visor and each other to prevent light penetration of the side viewing areas, thereby reducing low-angle solar glare.

The center eyeshade (fig. 2-28) is attached to the LEVA shell assembly over the shell thermal cover and can be lowered independently of the side eyeshade assemblies. When sufficiently lowered, the viewport door may be positioned as required to reduce solar glare. The viewport door is held in the desired position by a ratchet mechanism integral with the hinge assembly. The center eyeshade assembly cannot be independently lowered unless the protective visor and the sun visor are in the down positions.

2.3.5.10   Dual-Position Purge Valve

The purge valve (fig. 2-29) interfaces with the lower right exhaust (red) gas connector of the PGA. During contingency modes of EMU operation, the purge valve is operated in conjunction with the oxygen purge system (OPS) to complete an open-loop gas pressurization and ventilation system. When activated, the breathable gas flows from the oxygen purge bottle, through the PGA, and through the open purge valve to the outside atmosphere. Within the PGA, carbon dioxide is purged from the oronasal area and passes from the helmet down through the PGA ventilation distribution system to the purge valve. One of two purge flow-rate selections is available to the astronaut. High flow permits a normal 8.1-lb/hr flow of gas through the PGA with a 4.0-psia differential suit pressure.

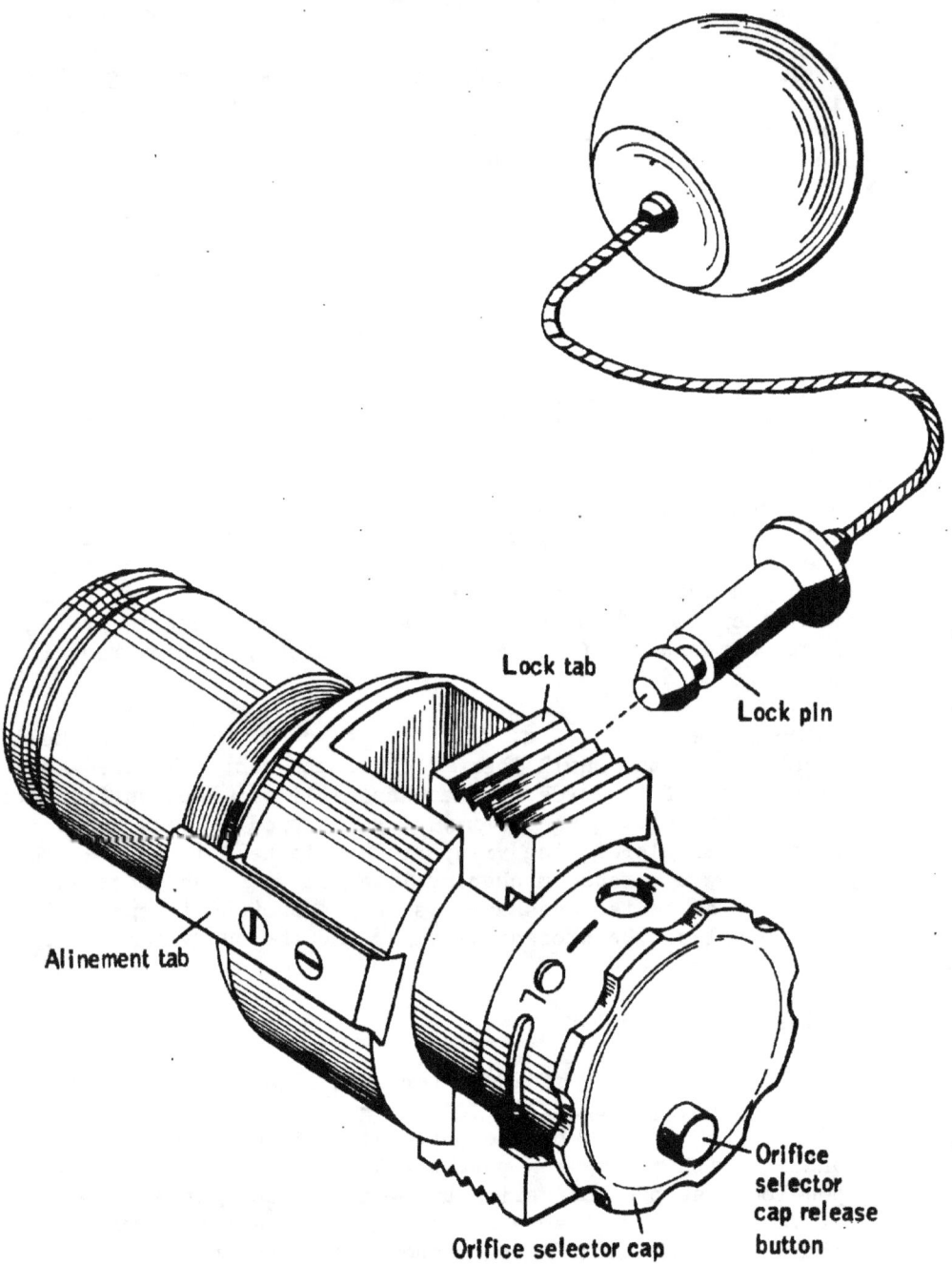

Figure 2-29.- Dual-position purge valve.

Low flow permits a normal 4.0-lb/hr flow. A lanyard unlocks the valve, and the valve is opened by depressing two lock tabs simultaneously. A rotating cap held by a release button provides the selection of low-flow-rate or high-flow-rate orifices.

2.3.5.11 Inflight Helmet Stowage Bag

The inflight helmet stowage bag (IHSB) (fig. 2-30) is used for temporary helmet stowage in the CM. It is constructed of a Teflon-coated Beta fabric and conforms to the helmet size.

2.3.5.12 LEVA Helmet Stowage Bag

The LEVA helmet stowage bag (fig. 2-31) consists of a formed polycarbonate base, shell assembly, and the necessary straps and components for attachment of the items to be stowed. The two-ply shell assembly and the polycarbonate base covering are made of Teflon-coated Beta cloth. Velcro strips are attached to the cover of the polycarbonate base to secure the LEVA stowage bag within the LM. The shell assembly is secured to the base assembly at the rear by two snaps and a tapered zipper closure (gusset) which draws the cover in snugly around the base. Additional security is provided around the bottom edge on each side of the gusset by Velcro strips. Polycabonat rings formed to the shape of the wrist disconnects are bonded to the polycarbonate base and provide stowage for the EV glove A polycarbonate retainer is also bonded to the base for stowag of the EMU maintenance kit. Straps with hook-and-pile fastene tape on the ends secure the EV gloves and EMU maintenance kit in position.

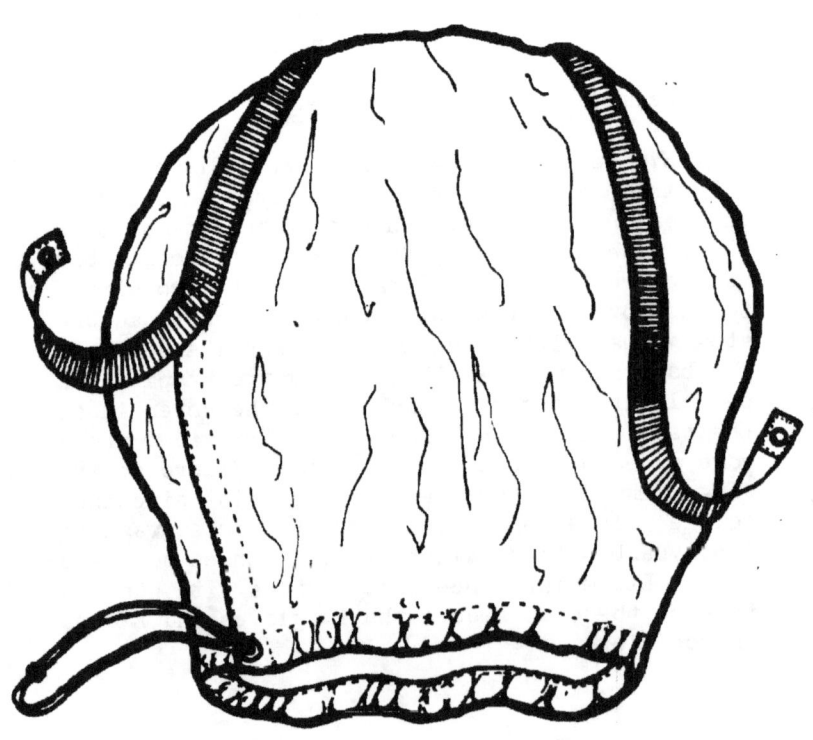

Figure 2-30.- Inflight helmet stowage bag.

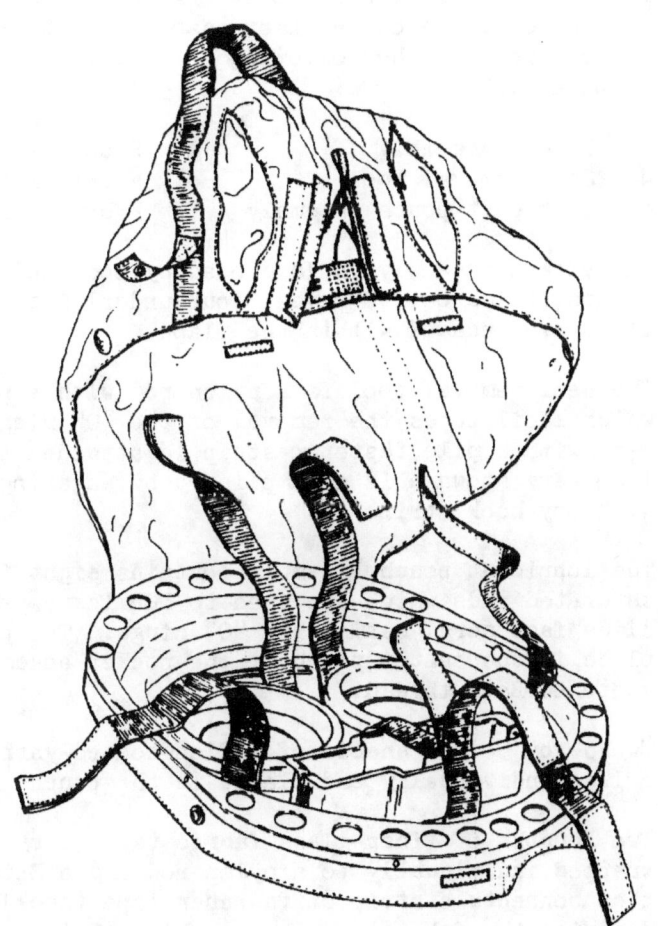

Figure 2-31.- LEVA helmet stowage bag.

2.3.5.13 EMU Maintenance Kit

The EMU maintenance kit (fig. 2-32) is a compact, lightweight assembly containing cleaning, replacement, and repair parts for inflight maintenance of the A7LB pressure garment assembly and the extravehicular visor assembly. The EMU maintenance kit consists of the following items each defined in subparagraphs:

a. Pocket assembly
b. Seal removal tool
c. Lubricant pouch assembly
d. Pouch assembly
e. Fabric repair patch
f. Fabric repair assembly

The pocket assembly, held closed by hook and pile fastener strips, folds out to reveal four underlying flaps. The six items are encased within the flaps.

The seal removal tool is a nylon rod with a preformed tip which facilitates the removal of the "O" ring seals. A lanyard with a pile fastener strip is attached to the tools. Temporary stowage is accomplished by engaging the pile strip with any hook strip.

The lubricant pouch assembly contains eight fluorinated, oil-saturated pads which are used to lubricate pressure sealing slide fasteners, seals, and "O" rings. The pads are held in place in the center pouch of the pocket assembly by whip-stitched Beta thread.

Two 5- by 5-inch sheets of Teflon-coated-yarn Beta cloth are rolled individually and placed in the pouch provided.

Two lengths of fiber-glass fabric tape (1 by 36 inches), wrapped individually to a nylon rod and a Beta-cord lanyard that connects a strip of fastener tape (hook) to the rod, comprise the fabric repair assembly. This assembly is stowed in a pocket provided in the EMU maintenance kit. The tape may be employed to complete small repairs to layers of the ITMG or CLA or used in conjunction with the Teflon-coated Beta cloth when repairs to abraded, cut, or torn areas of the ITMG or CLA are required.

The pouch assembly consists of six transparent, heat-sealed pouches. Each pouch is clearly labeled as to its contents. The entire pouch assembly is attached to the pocket assembly by snap fasteners.

Amendment 2
11/5/71

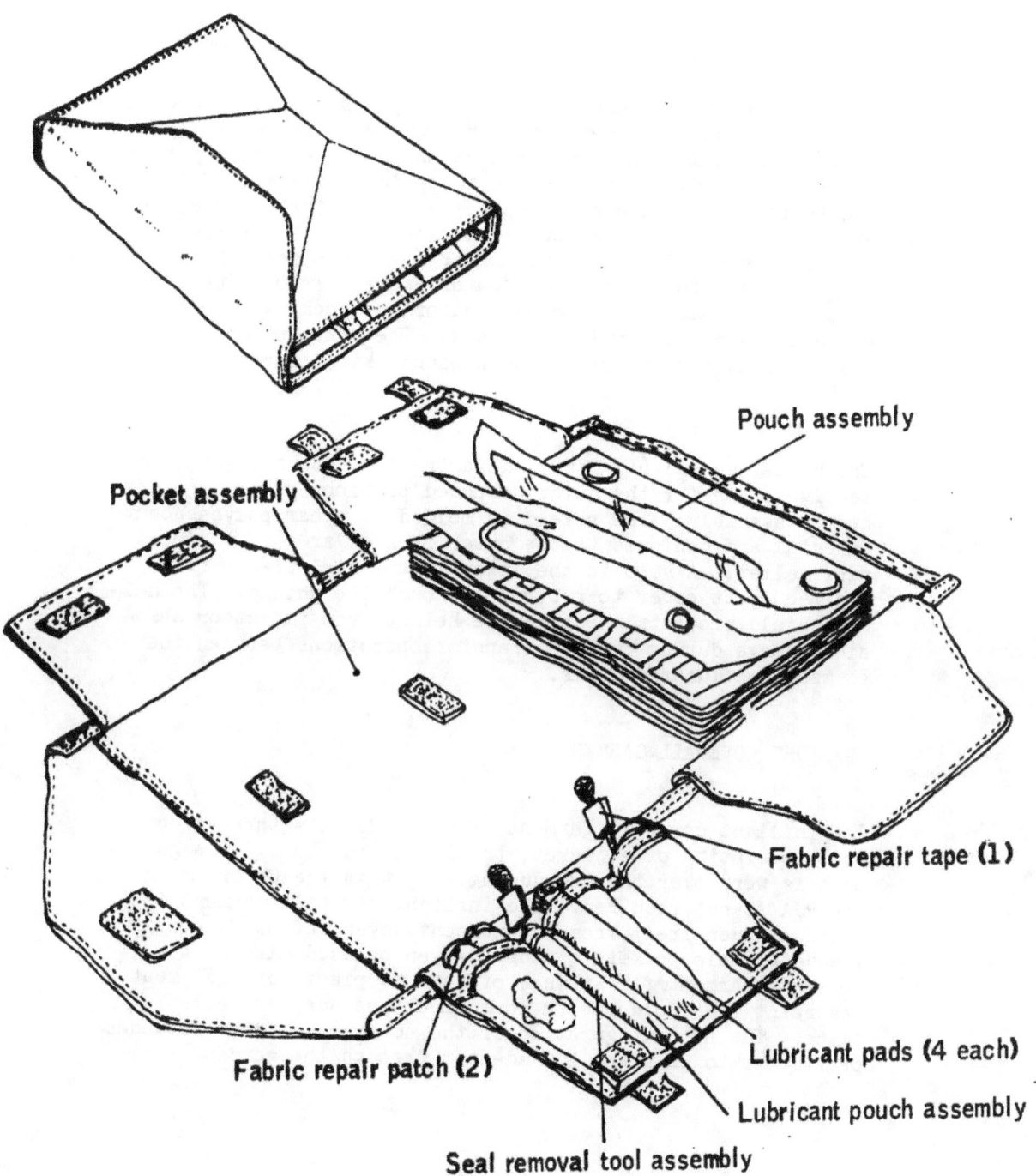

Figure 2-32.- EMU maintenance kit.

The first pouch contains three repair patches made of pressure-sensitive tape. The second pouch contains five pockets of sealant which are used in conjunction with the repair patches to seal accidental punctures in the primary bladder of the PCG. The third pouch contains a replacement seal for a large wrist disconnect. The fourth pouch contains three compartments, one for a spare PRV "O" ring, one for a spare feedport "O" ring, and one for a spare gas/water connector "O" ring. The fifth pouch contains three applicator pad pockets each of which contains two applicator pads. The sixth pouch contains instructions for use of the maintenance kit contents.

2.3.5.14 Helmet Shield

The helmet shield (fig. 2-5) is a transparent, slip-on, protective cover for the outer, exposed portions of the pressure helmet assembly. The shield is molded of clear polycarbonate material and conforms to the outer frontal area of the pressure helmet. A hole in the lower left facial area permits the feed-port cover to protrude through the shield. The helmet shield protects the pressure helmet from impact or abrasion damage during crewman transfer operations between the command and lunar modules.

2.4 INFLIGHT COVERALL GARMENT

The inflight coverall garment (fig. 2-33) is a three-piece suit consisting of a jacket, trousers, and boots. The garment is worn over the CWG during flight in the CM or LM when the PGA is not required. The inflight coverall garment is fabricated entirely from 100-percent woven Teflon fabric. The detachable pockets of the PGA can be used also on the coverall garment for stowage of various pieces of equipment. Restraint tabs hold the CM communications adapter cable in place. The LM configuration of the coverall garment includes provisions to pass the LCG adapter through the garment.

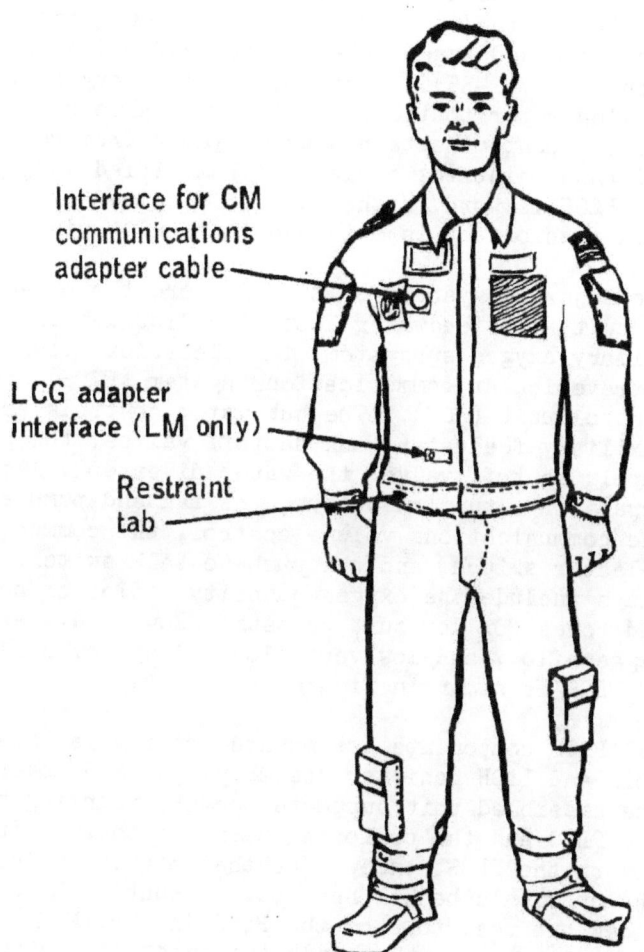

Figure 2-33.- Inflight coverall garment.

2.5 PORTABLE LIFE SUPPORT SYSTEM

The PLSS (fig. 2-34) provides life support for EV EMU activity, including expendables for metabolic consumption, communications, telemetry, operating controls, and displays. Although the -7 PLSS used for Apollo 15 and subsequent missions is similar to the -6 PLSS used on previous missions, the -7 PLSS has increased expendables capacity for longer duration missions (fig. 2-35). The PLSS supplies oxygen to the PGA and cooling water to the LCG. The PLSS also removes solid and gaseous contaminants and water vapor from returning oxygen and thus maintains a clean, dehumidified supply of oxygen. The PLSS is worn on the back of a suited astronaut in knapsack fashion and is attached to the PGA with harnesses.

The major subsystems of the PLSS are the oxygen ventilation circuit, the feedwater loop, the liquid-transport loop, the primary oxygen subsystem, the electrical power subsystem, the extravehicular communications system (EVCS), and the remote control unit (RCU). The subsystem controls are the main and auxiliary feedwater tank shutoff valves, the primary oxygen supply shutoff valve, the water diverter valve, the gas-separator actuation button, the fan and pump actuation switches, the communications volume control, the communications mode-selector switch, and the push-to-talk switch. Subsystem displays include the oxygen quantity indicator and warning flags and tones for low suit pressure, low feedwater pressure, high oxygen flow, and low vent flow. A system schematic of the -7 PLSS is shown in figure 2-36.

All PLSS components are mounted on the main feedwater reservoir and LiOH canister assembly. A hard cover fitted over the assembled unit supports the OPS mounting plate on top of the PLSS and the conformal pads. A thermal insulation jacket covers the PLSS, except for that portion which is exposed to the crewman's back. Hard-point mounting holes in the PLSS sides are used to stow the PLSS in the LM during flight and may be used to mount the buddy secondary life support system (BSLSS) during EVA.

2.5.1 Oxygen Ventilating Circuit

The oxygen ventilating circuit supplies fresh, cooled oxygen at 3.5 to 4.0 psia through the PGA. A fan motor assembly forces the oxygen into the PGA at a flow rate of 5.5 acfm with a minimum pressure rise of 1.5 inches of water. Suit

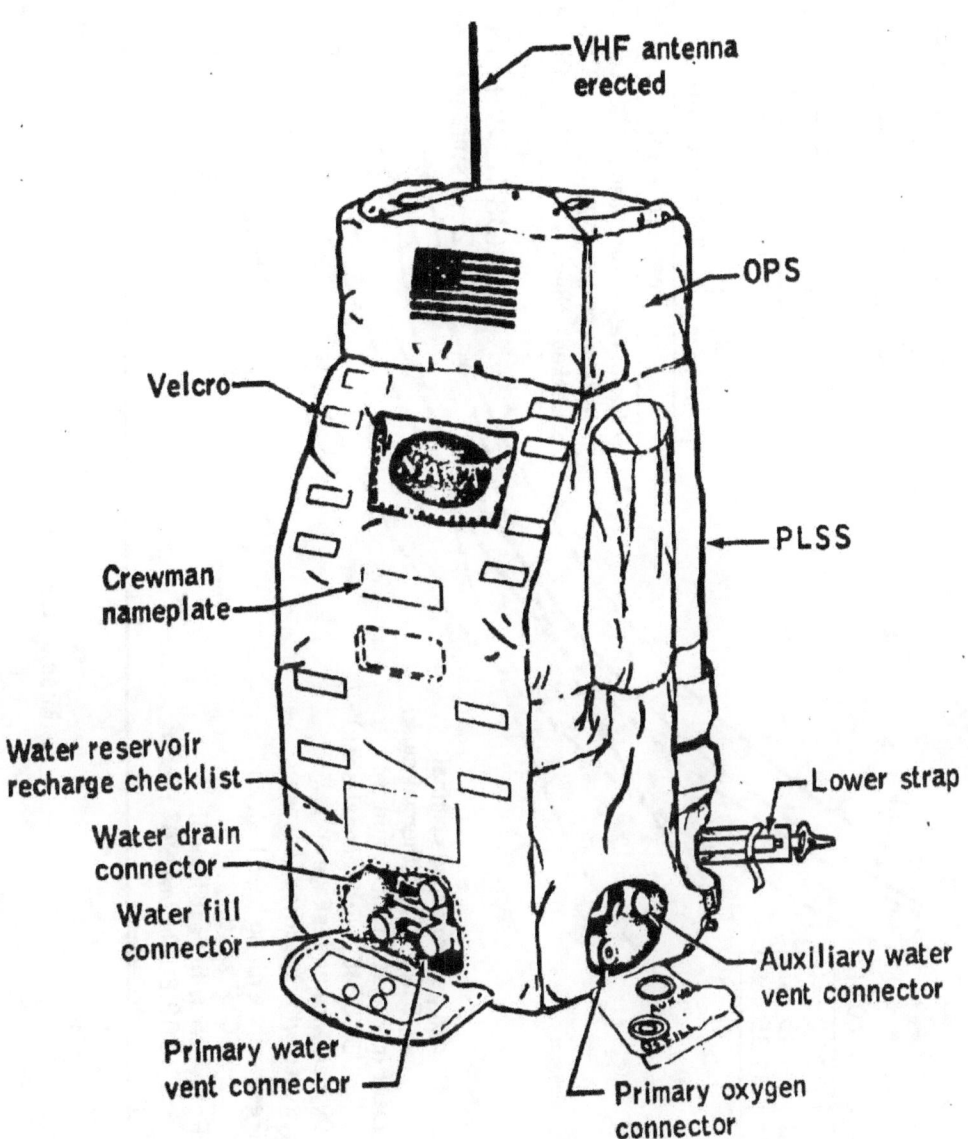

Figure 2-34.- Portable life support system.

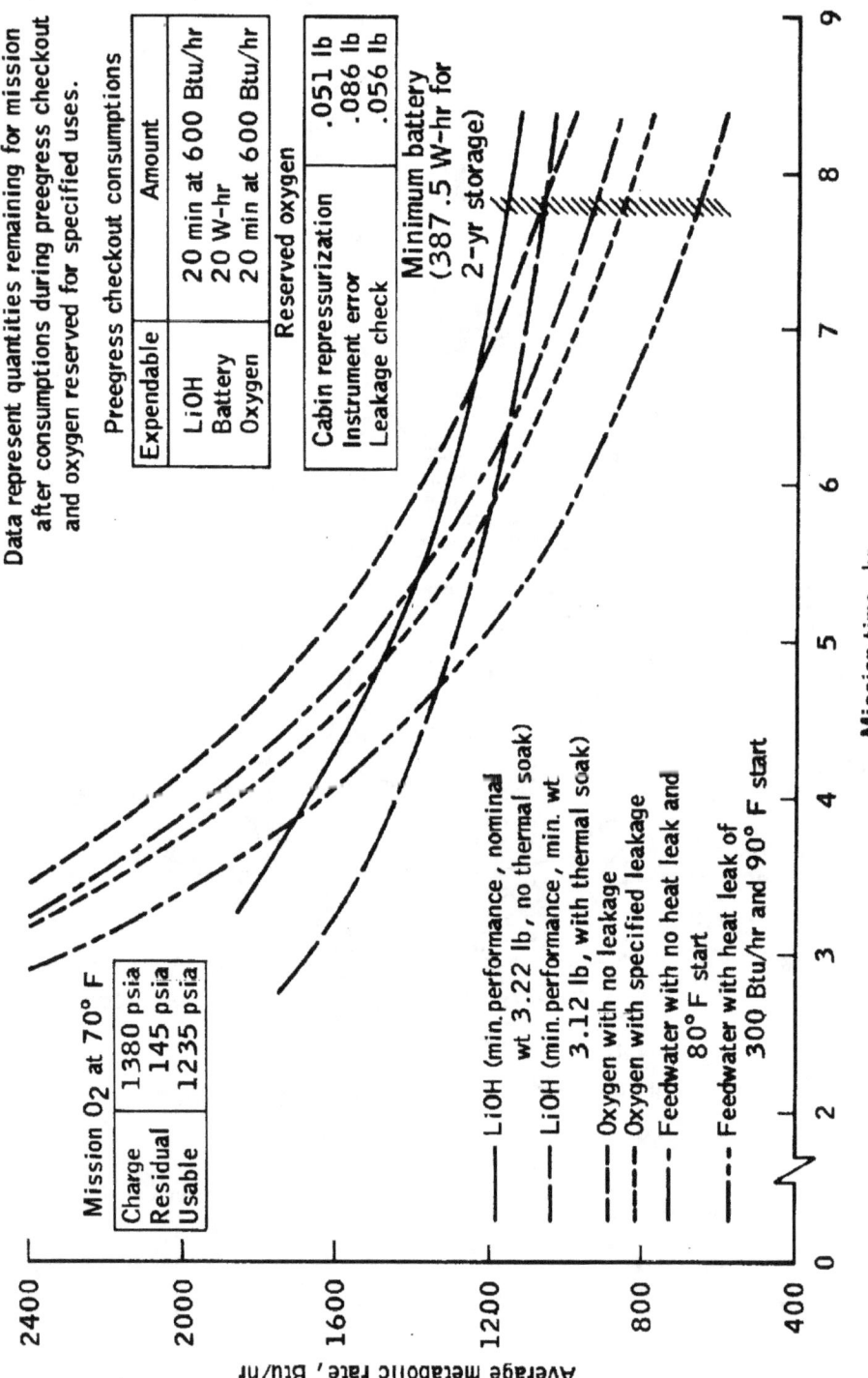

Figure 2-35.- Duration of -7 PLSS expendables.

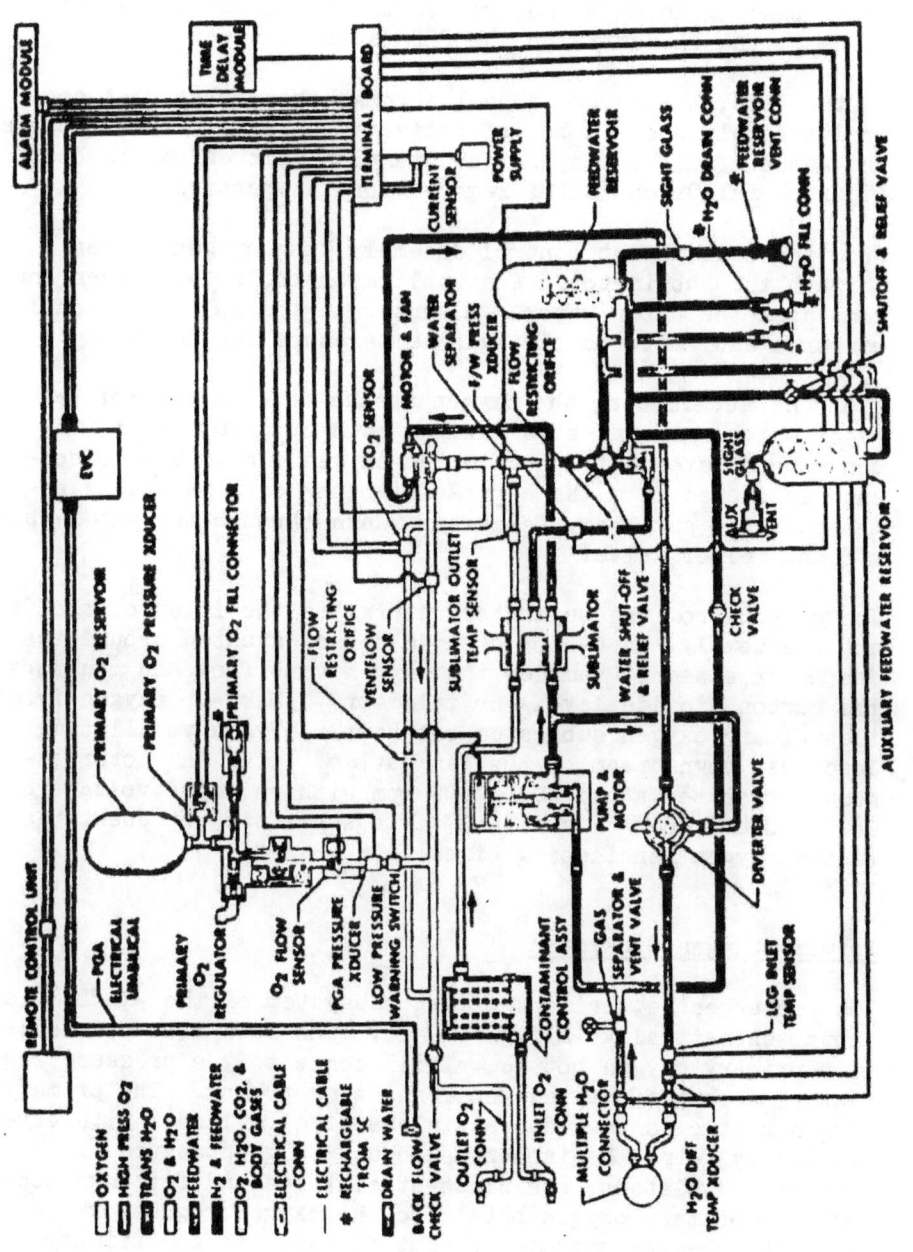

Figure 2-36.- System schematic of the -7 PLSS.

inlet dewpoint temperature is 50° F (or below), and suit inlet oxygen temperature is approximately 77° F (nominal). After passing through the suit vent system, the oxygen returns to the PLSS through the PLSS inlet connector.

In the PLSS, the oxygen passes through the contaminant control assembly where a bed of activated charcoal removes odors and a bed of lithium hydroxide granules removes carbon dioxide. A peripheral Orlon filter removes foreign particles.

From the contaminant control assembly, the oxygen passes through the sublimator. The sublimator cools the oxygen and condenses the water vapor. A sensor at the sublimator outlet measures sublimator outlet gas temperature for telemetry.

From the sublimator, the oxygen passes through a water separator which removes, at a maximum rate of 0.508 lb/hr, the condensate water entrapped in the oxygen flow. The condensate is ducted from the separator to the outer sections of the main and auxiliary feedwater tanks through the water shutoff and relief valve.

The oxygen from the separator returns to the inlet of the fan motor assembly. A carbon dioxide sensor shunted around the fan motor assembly samples the oxygen vent flow and monitors the carbon dioxide level for telemetry. Make-up oxygen from the primary oxygen subsystem enters the oxygen ventilating loop just downstream of the fan outlet. (The fan motor assembly operates at 18 600 ± 600 rpm with an input voltage of 16.8 ± 0.8 V dc.) Figure 2-37 is a schematic representation of the oxygen ventilating circuit.

2.5.2   Primary Oxygen Subsystem

The rechargeable, primary oxygen subsystem of the -7 PLSS is shown schematically in figure 2-38. The subsystem consists of a primary oxygen bottle, a fill connector, a pressure regulator, a shutoff valve, and connecting tubing. The primary oxygen bottle is a welded stainless-steel cylinder with cryogenically formed hemispherical ends. High-pressure, corrosion-resistant, stainless-steel tubes and fittings connect the primary oxygen bottle to the oxygen regulator assembly. The crewman actuates a shutoff valve to the primary oxygen regulator assembly by an operating lever located at the lower-right-front corner of the PLSS. When the PLSS is not in use or when the primary oxygen subsystem is being charged, the oxygen shutoff valve is closed.

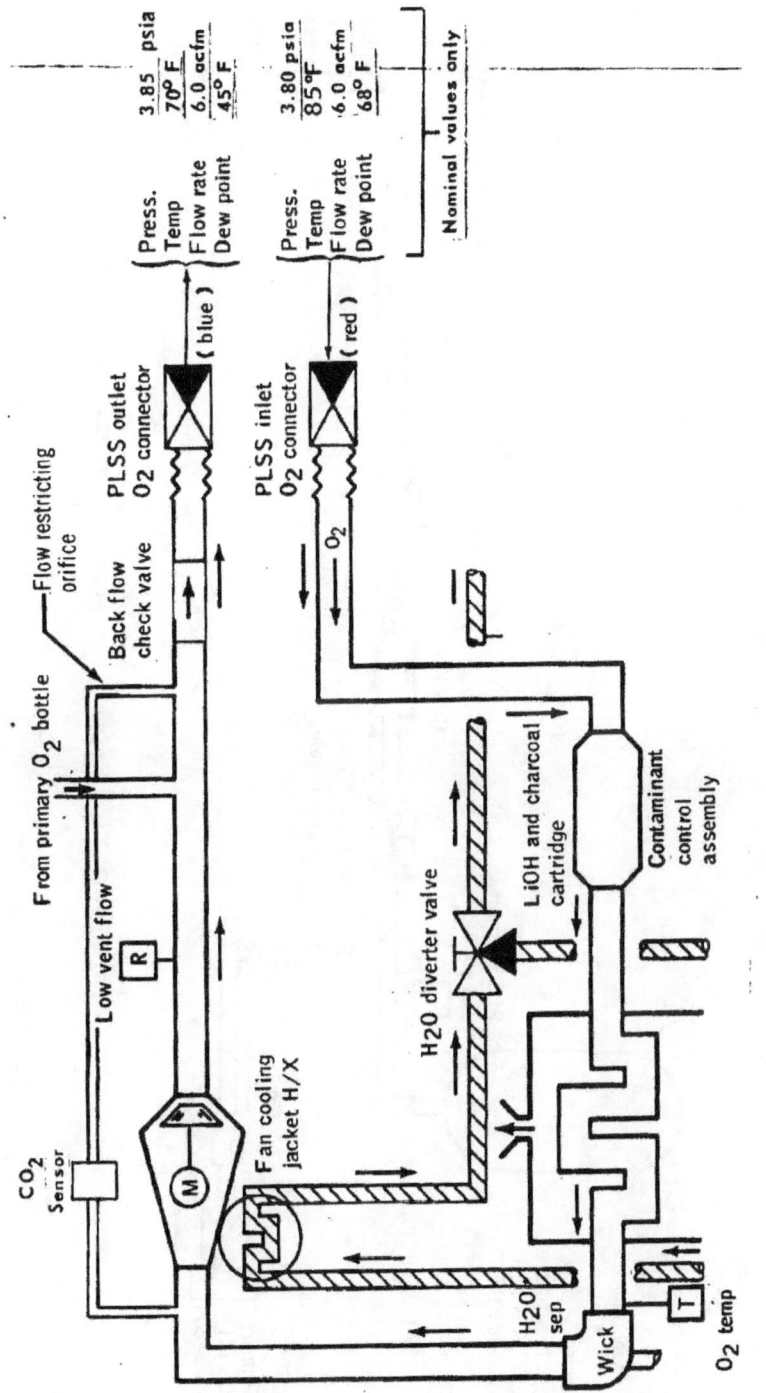

Figure 2-37.- Oxygen ventilating circuit.

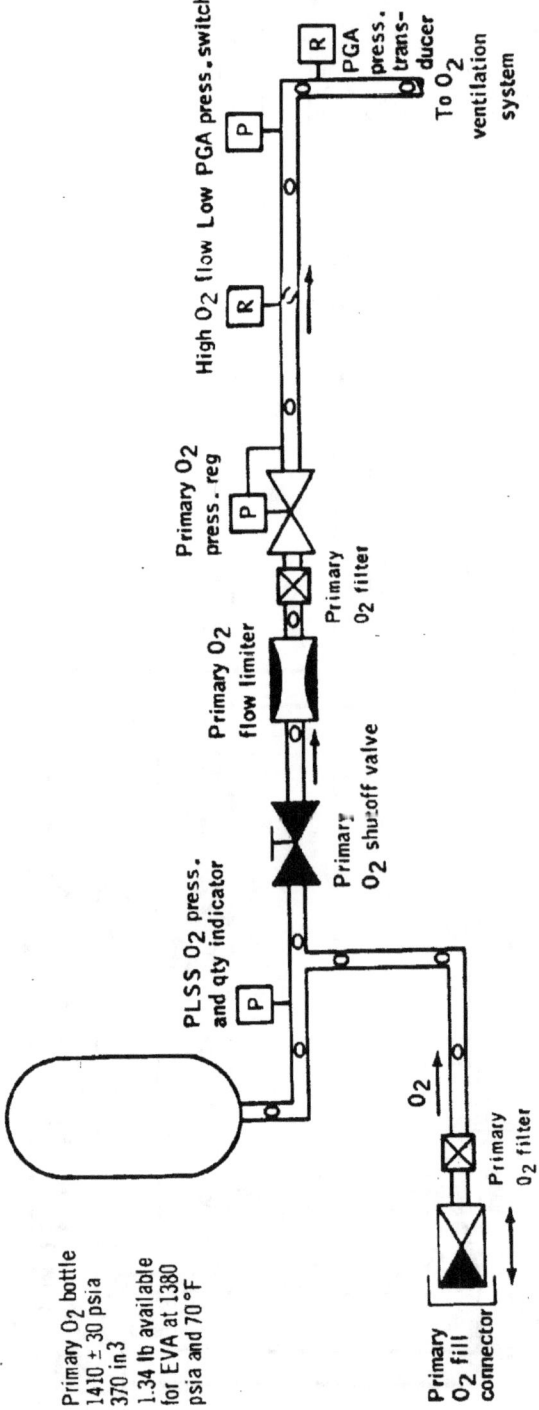

Figure 2-38.- Primary oxygen subsystem.

The initial ground charge and the LM recharge pressure for the first four recharges is 1410 ± 30 psia. Recharge pressure for the fifth recharge is 1310 psia minimum. The charging process (except for the fifth recharge) gives a minimum of 1.340 pounds of usable oxygen for EVA at 1380 psia and 70° F. This oxygen supply is ample for a 5-hour EVA at 1200-Btu/hr metabolic load plus 300 Btu/hr of EMU leakage.

Make-up oxygen flows from the primary oxygen bottle through the shutoff valve and regulator to the oxygen ventilating circuit. The regulator provides a pressure of 3.85 ± 0.15 psia to the vent circuit. An orifice limits the flow to a maximum of 4.0 pounds per hour at 70° F with a supply pressure of 1500 psia, thereby protecting the PGA from overpressurization if the regulator fails open. A primary oxygen pressure transducer at the oxygen bottle outlet provides electrical signals to the RCU oxygen quantity indicator and to the PLSS telemetry system. If oxygen flow exceeds 0.50 to 0.65 pound per hour, an oxygen flow sensor downstream of the regulator gives an audible tone until the flow decreases to 0.50 to 0.65 pound per hour (a continuous high flow of 0.50 to 0.65 pound per hour for 5 seconds is needed to cause actuation). Two additional pressure transducers in the primary oxygen subsystem are used to monitor PGA pressure. One is used for telemetry monitoring, and the other activates an audible warning tone when pressure drops below 3.10 to 3.40 psid. The primary oxygen subsystem is recharged through a leak-proof, self-sealing, quick-disconnect fill connector.

2.5.3 Liquid Transport Loop

The recirculating liquid transport loop provides thermal control for the crewman by dissipating heat through the sublimator. Warm transport water from the LCG enters the PLSS through the MWC. The water then passes through a gas separator which can entrain a minimum of 30 acc of gas. Should cooling performance degrade because of additional gas, the crewmen can vent the trap manually to ambient and ready it for further entrapment. From the separator, the transport water enters the pump which forces the water through the sublimator for cooling. The pump provides a minimum flow of 4.0 pounds per minute with a pressure rise of 1.9 psi across the inlet and outlet portions of the PLSS MWC. The cooled water from the sublimator passes through the fan motor cooling jacket and then through the diverter valve and out of the MWC.

The crewman regulates coolant flow with the diverter valve. In the minimum position, most of the flow is diverted past the sublimator. In the maximum position, all of the flow from the LCG passes through the sublimator. The intermediate position provides midrange cooling. The liquid transport loop is interconnected to the feedwater loop by a check valve which permits make-up water to enter the transport loop upstream of the pump.

A differential temperature transducer senses the differential temperature of LCG water entering and leaving the PLSS, and a temperature transducer senses LCG inlet temperature. Both transducers provide electrical signals for telemetry.

A schematic of the liquid transport loop is shown in figure 2-39.

2.5.4 Feedwater Loop

The feedwater loop is shown schematically in figure 2-40. This loop contains a primary feedwater reservoir and an auxiliary feedwater reservoir. The reservoirs supply water to the porous plate of the sublimator and collect condensation supplied by the water separator.

Each reservoir is a bladder-type rechargeable tank. Minimum capacities are 8.40 pounds of water for the primary reservoir and 3.06 pounds of water for the auxiliary reservoir. Feedwater from both reservoirs flows through a manually operated shutoff and relief valve. This valve, when in the off position, acts as a relief valve to prevent overpressurization of the feedwater reservoir. Feedwater then enters the porous plate of the sublimator. The feedwater forms an ice layer on the surface of the porous plate which is exposed to vacuum. Heat from the liquid transport loop and oxygen ventilating circuit is conducted to the porous plate and is dissipated by sublimation of the ice layer. A flow-limiting orifice between the shutoff and relief valve and the sublimator prevents excess water spillage from the sublimator porous plate during startup or during a possible sublimator breakthrough (a condition in which ice fails to form on the surface of the porous plate). A separate shutoff and relief valve isolates the auxiliary feedwater reservoir from the primary feedwater reservoir during normal operation. If the primary feedwater supply is depleted during EVA, the crewman can open the auxiliary reservoir shutoff and relief valve to provide additional

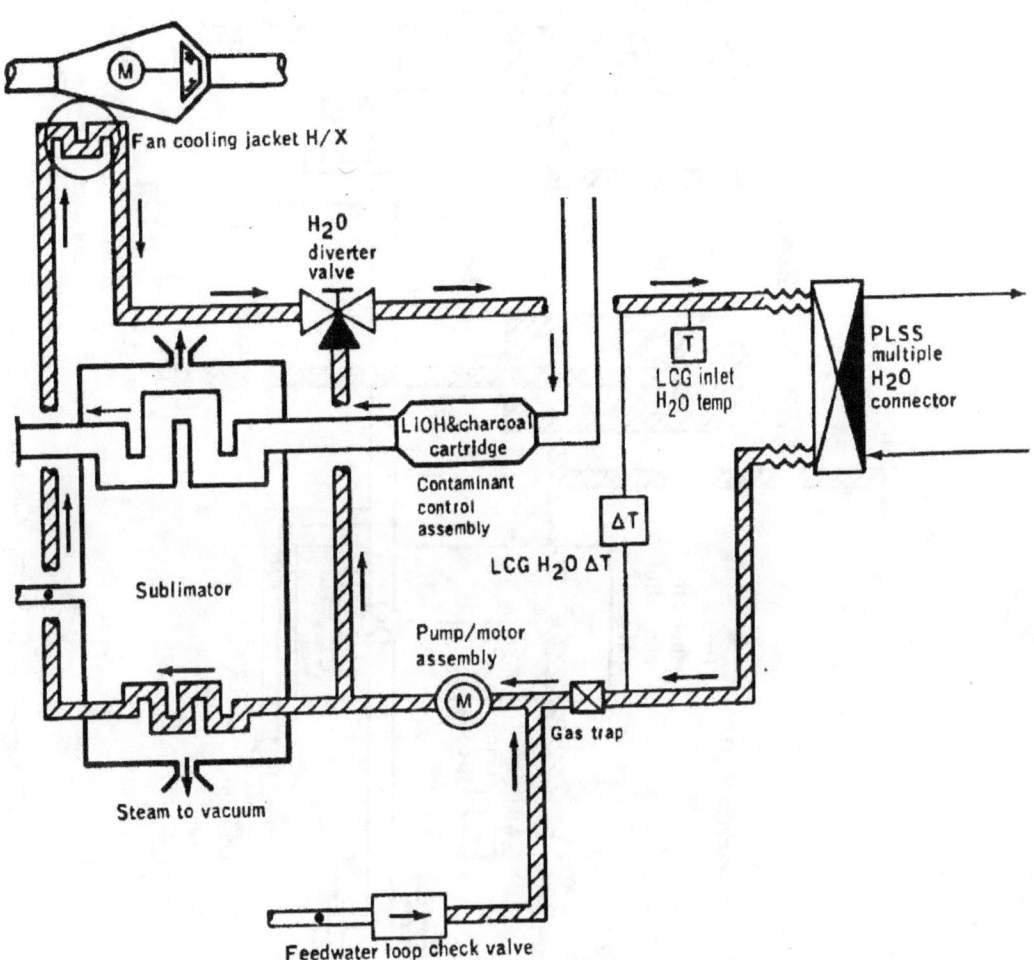

Figure 2-39.- Liquid transport loop.

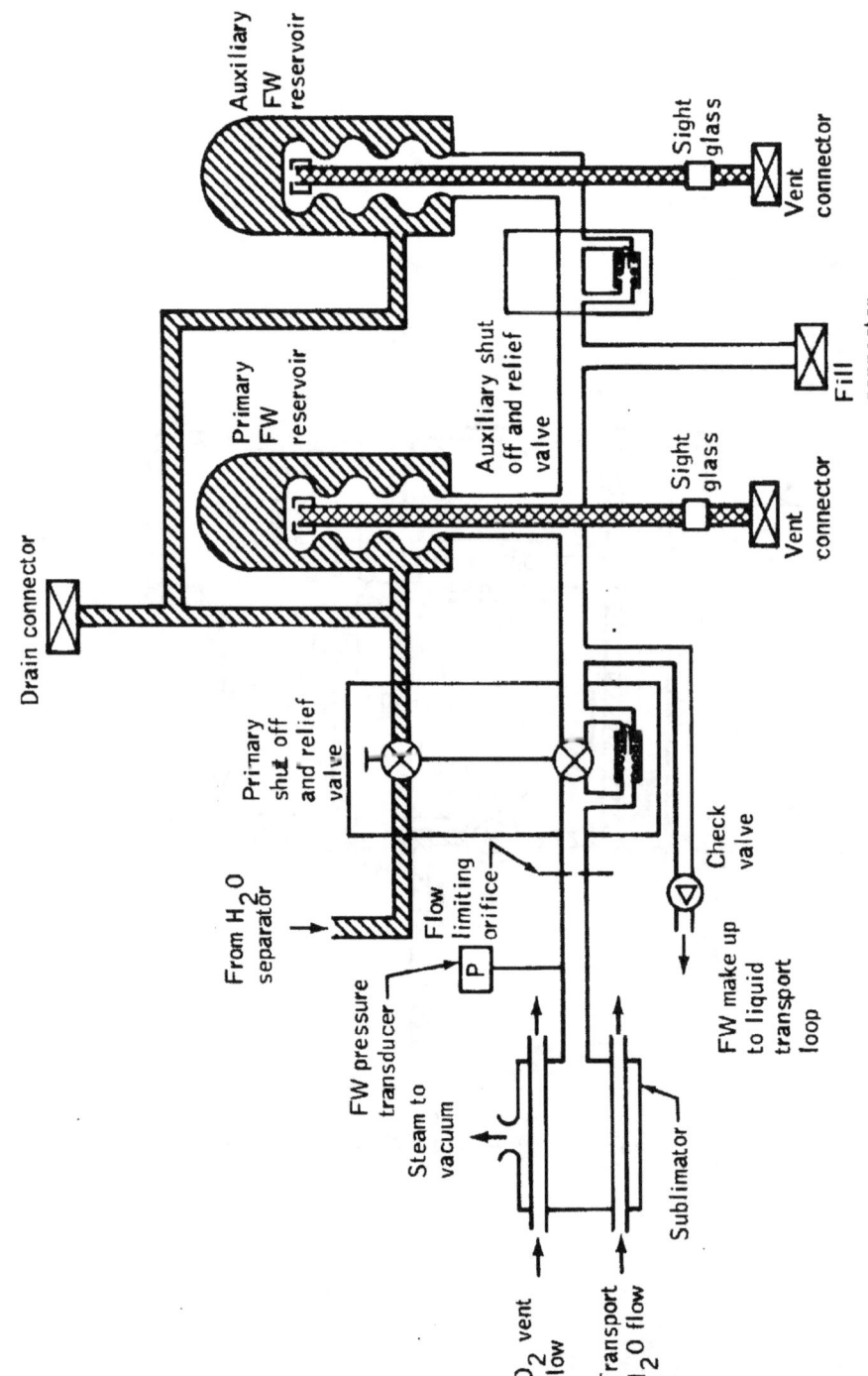

Figure 2-40.- PLSS feedwater loop.

cooling. Both the primary and auxiliary feedwater shutoff and relief valves are actuated by handles at the lower-right-front corner of the PLSS. The feedwater reservoirs also provide make-up water to the liquid transport loop via a check valve.

Oxygen ventilating loop pressure forces the condensate from the water separator into the space between the reservoir housings and the bladders of both feedwater reservoirs. This action causes a pressure of 3.3 psid on the feedwater bladder.

The feedwater reservoirs are recharged and drained through fill and drain connectors attached to both sides of the bladders. Recharge and drainage are performed simultaneously. Each bladder contains a vent line with a vent connector. During recharge, the vent connector is connected to a vacuum line to remove entrapped gas and assure a full charge.

A feedwater pressure transducer just upstream of the sublimator provides telemetry monitoring to identify sublimator breakthrough or feedwater depletion. The transducer also contains a switch which actuates an audible warning and the low feedwater pressure warning flag on the RCU if feedwater pressure drops to 1.2 to 1.7 psia.

2.5.5 Electrical Power Subsystem

The electrical power subsystem provides dc electrical power through appropriate connectors to the fan motor assembly, the pump motor assembly, and for communications and instrumentation. A 16.8 ± 0.8-V dc, 11-cell, silver-zinc alkaline battery supplies the power.

The minimum capacity of the -7 PLSS power supply is 387.5 watt hours which is for a battery shelf life of 2 years.

The sliding pin locking device, shown in figure 2-41, holds the battery in place. Between extravehicular activities, a crewman can release this device to replace the battery.

2-94   CSD-A-789-(1) REV V

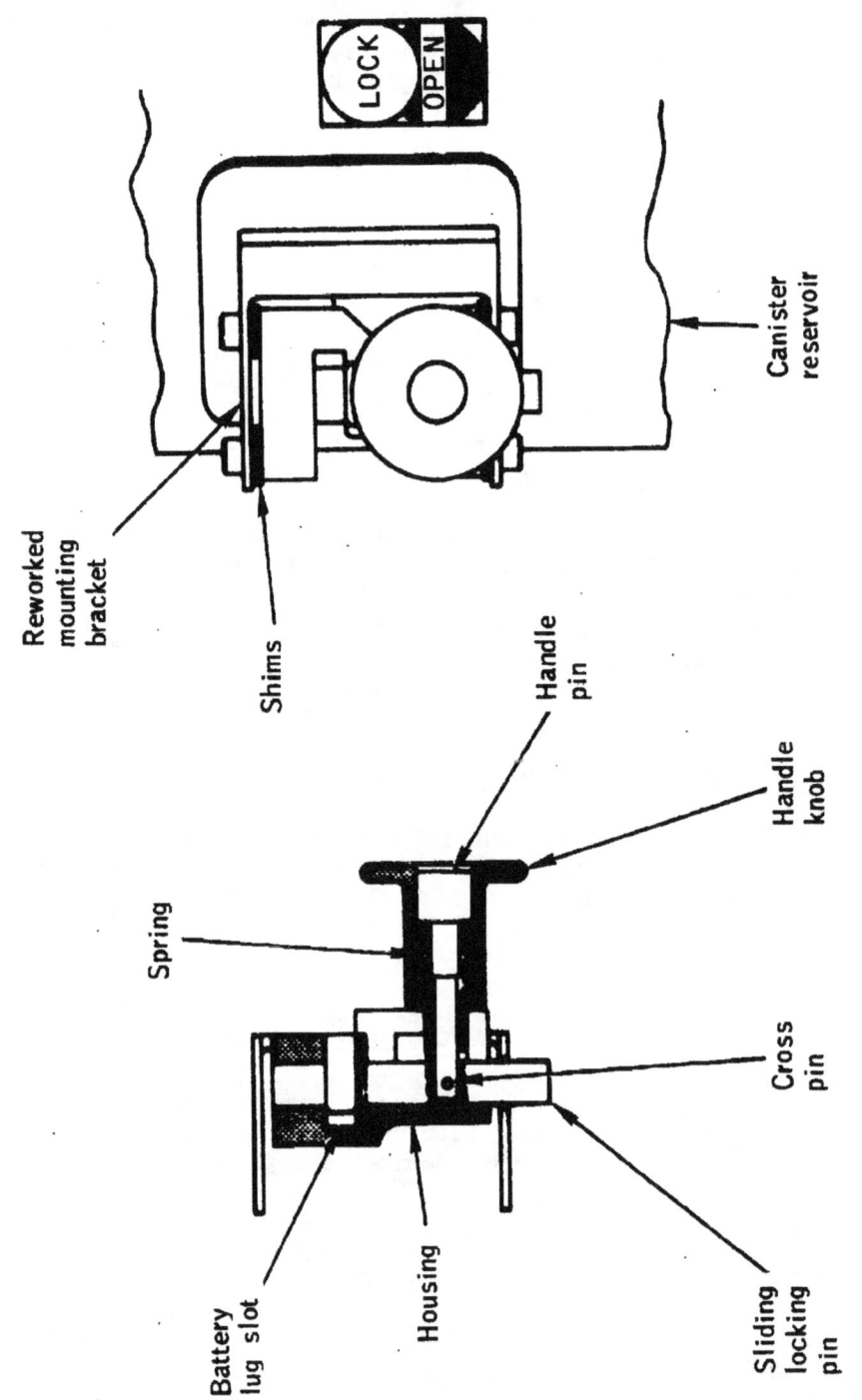

Figure 2-41.- Battery locking device.

Electrical power requirements are as follows.

	Maximum, watt	Nominal, watt
Pump	10.0	8.4
Fan	32.5	21.8
EVCS	12.8	10.9

Current limiters protect selected electric circuits against overcurrents which could cause fires. These limiters pass transient current in excess of a normal load but open at sustained overload. Table 2-X lists current limiter ratings. Transducers provide signals for telemetry of battery current and voltage.

2.5.6 Extravehicular Communications System

The EVCS (fig. 2-42) provides the following basic capabiliti

a. Simultaneous and continuous telemetry from two extravehicular crewmen

b. Duplex voice communications between earth and one or both of the two extravehicular crewmen

c. Uninterruptable voice communications between the crewmen

d. Thirty telemetry channels, 30 by 1-1/2 pam, per each extravehicular communicator (EVC) with 26 channels available for status information

e. Separate subcarrier frequencies for continuously monitoring each crewman's ECG during EVA

f. An audible alarm for 10 ± 2 seconds in the event of an unsafe condition (if the EVC mode-selector switch position is changed and the unsafe condition still exists, the warning tone will come on again for 10 ± 2 seconds.)

The EVCS consists of two extravehicular communicators (EVC-1 and EVC-2) which are an integral part of the PLSS. The EVC-1 consists of two amplitude modulation (AM) transmitters, two

TABLE 2-X.- PLSS/EVCS CURRENT LIMITER RATINGS

Component	Current ratings of -7 configuration, A
Fan	22-gage wire — current protection is not provided[a]
Pump	22-gage wire — current protection is not provided[a]
ECG	1/4 (with series 32.4- to 39.2-ohm, 1/2-watt resistor)
Left microphone	1/8 (with series 32.4- to 39.2-ohm, 1/2-watt resistor)
Right microphone	1/8 (with series 32.4- to 39.2-ohm, 1/2-watt resistor)
Vent flow sensor	1/16
Time delay module	1/16
High O_2 flow sensor	None (unit has built-in current limiter)
EVC (dual-primary mode voltage regulator)	2
EVC (secondary mode voltage regulator)	2
EVC telemetry	1
Alarm module	1/2
Voltage regulators	3/4
Time delay module (for high O_2 flow sensor)	1/16
EVC warning tone generator	1/16

[a]The maximum overload current of 22-gage copper wire is 40 amps.

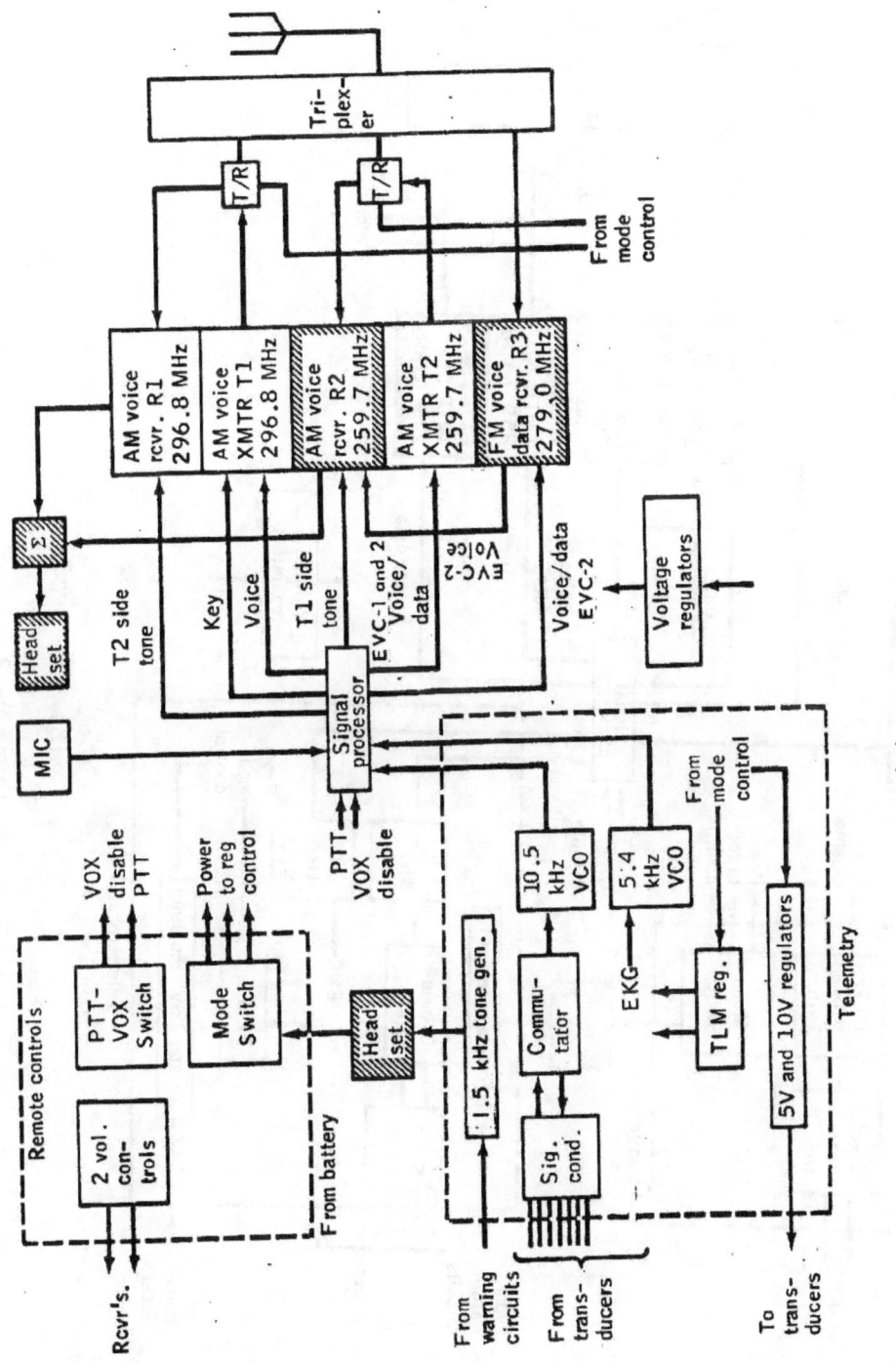

Figure 2-42.- Extravehicular communications system.

(a) The EVC-1.

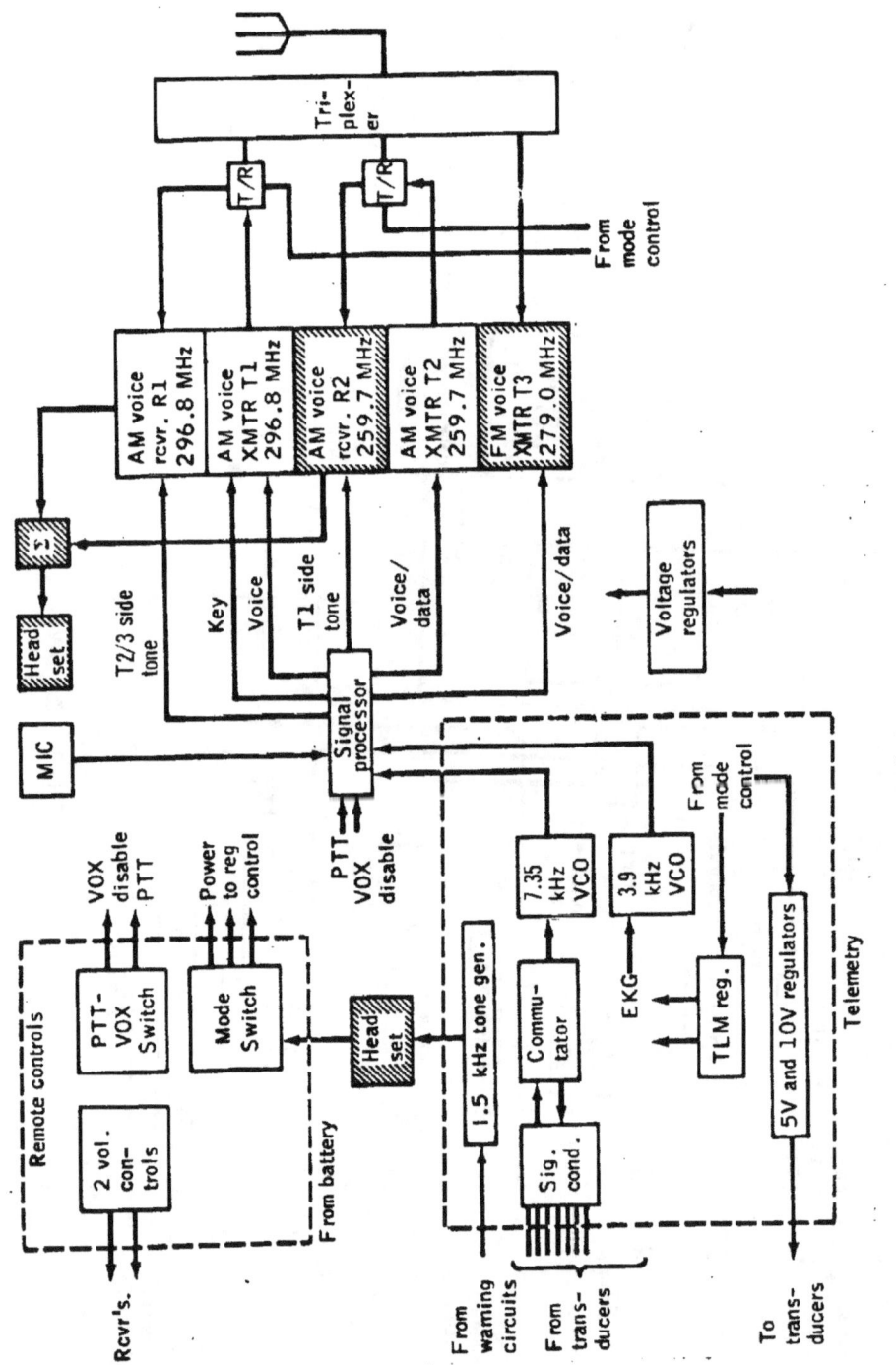

(b) The EVC-2.

Figure 2-42.- Concluded.

AM receivers, one frequency modulation (FM) receiver, signal-conditioning circuits, a telemetry system, a warning system, and other components required for system operation. The EVC-2 is similar to the EVC-1 except that the EVC-2 has an FM transmitter instead of an FM receiver.

Each EVC can be controlled manually by a four-position switch for each of the following modes of operation.

a. Off (O)
b. Dual (AR)
c. Primary (A)
d. Secondary (B)

The dual mode is the normal operating position of the switch. In this mode, the EVC-2 transmits a 0.3- to 2.3-kHz voice signal and two interrange instrument group (IRIG) subcarriers (3.9 and 7.35 kHz) via a 279-MHz FM transmitter. The transmitter has an unmodulated output in excess or 500 mW. The composite signal from the EVC-2 is received at EVC-1, mixed with an additional 0.3- to 2.3-kHz voice signal and two additional IRIG subcarriers (5.4 and 10.5 kHz), and transmitted to the LM on a 259.7-MHz AM link. The composite signal of two voice and four subcarriers is then relayed from the LM to the earth via S-band. The EVC-2 also receives EVC-1 output (which includes the original EVC-2 transmission) on a 259.7-MHz receiver; thus, a duplex link between the two EV crewmembers is established. Communications signals are transmitted from the earth to the LM via S-band and are then relayed to both astronauts on the 296.8-MHz AM link.

The outputs of the FM and AM receivers are summed with an attenuated input voice signal and applied to the earphones. The audio output levels of both receivers are individually controlled by separate volume controls located in the RCU affixed to the chest of the PGA. The input voice signal is attenuated 10 dB to provide a sidetone for voice level regulation.

The dual mode provides uninterruptable duplex voice communications between the crewmembers and the LM/earth linkup plus simultaneous telemetry from each crewmember via relay through EVC-1.

In the event of a malfunction in the dual mode, the system is backed up by the primary- and secondary-mode positions. (Note that both crewmen should never be in the primary or secondary modes simultaneously. Severe distortion and interference will occur, and communications will be temporarily lost.)

In the primary and secondary modes, duplex voice communication is maintained between the two crewmen and the LM. The secondary mode, however, has no telemetry capability. Also, the secondary-mode transmitter is inoperative unless activated by the voice-operated switch or the manual switch. The transmitter is continually operative in the dual and primary modes.

The telemetry unit contains a warbling 1.5-kHz warning tone. Any one of four problems (high oxygen flow, low vent flow, low PGA pressure, or low feedwater pressure) will key the tone and alert the astronaut to check the remote control unit for a visual indication of the problem area to be investigated. The operation of the warning system is independent of mode selection.

Each telemetry system can accommodate up to 26 commutator channels (table 2-XI) at 1-1/2 samples per second and one ECG channel and provides a data accuracy of 2 percent root mean square.

2.5.7 <u>Remote Control Unit</u>

The RCU (fig. 2-43) is a chest-mounted instrumentation and control unit which provides the crewman with easy access to certain PLSS/EVCS controls and displays. Controls include a fan switch, a pump switch, a communications mode-selector switch, a push-to-talk switch, and two communications volume control knobs. Displays include an oxygen quantity indicator and four active status indicators (warning flags). A fifth status indicator is provided, but is not presently used. The status indicators are illuminated by beta particle capsules requiring no electricity. Any one of four problems (high oxygen flow, low vent flow, low PGA pressure, or low feedwater pressure) will cause a cylinder to rotate and reveal the illuminated warning symbol underneath. Simultaneously, the warble tone in the EVCS is activated to alert the crewman

TABLE 2-XI.- PLSS/EVCS COMMUNICATIONS TELEMETRY CHARACTERISTICS

Measurement title	Instrumentation range	Discriminator output range, V dc	Commutator channels
Zero calibration	0 V dc	0	1
Full-scale calibration	5 V dc	5	2
PGA pressure	2.5 to 5.0 psid	0 to 5	3, 21, 24, 27
Feedwater pressure	0 to 5.0 psia	0 to 5	4, 15, 22, 26
Battery current	0 to 10 amps	0 to 5	5, 11
Battery voltage	15.5 to 20.5 V dc	0 to 5	6, 20
Water difference temperature	0° to 15° F	0 to 5	8, 19
LCG inlet temperature	40° to 90° F	3.13 to 1.86	9, 17
Sublimator gas outlet temperature	40° to 90° F	3.13 to 1.86	10, 16
Primary oxygen pressure	0 to 1110 psia	0 to 5	12, 13, 23, 28
Carbon dioxide partial pressure	0.1 to 30 mm Hg	0 to 5	7, 14, 18, 25
Synchronization	- -	(Double width pulse)	29, 30

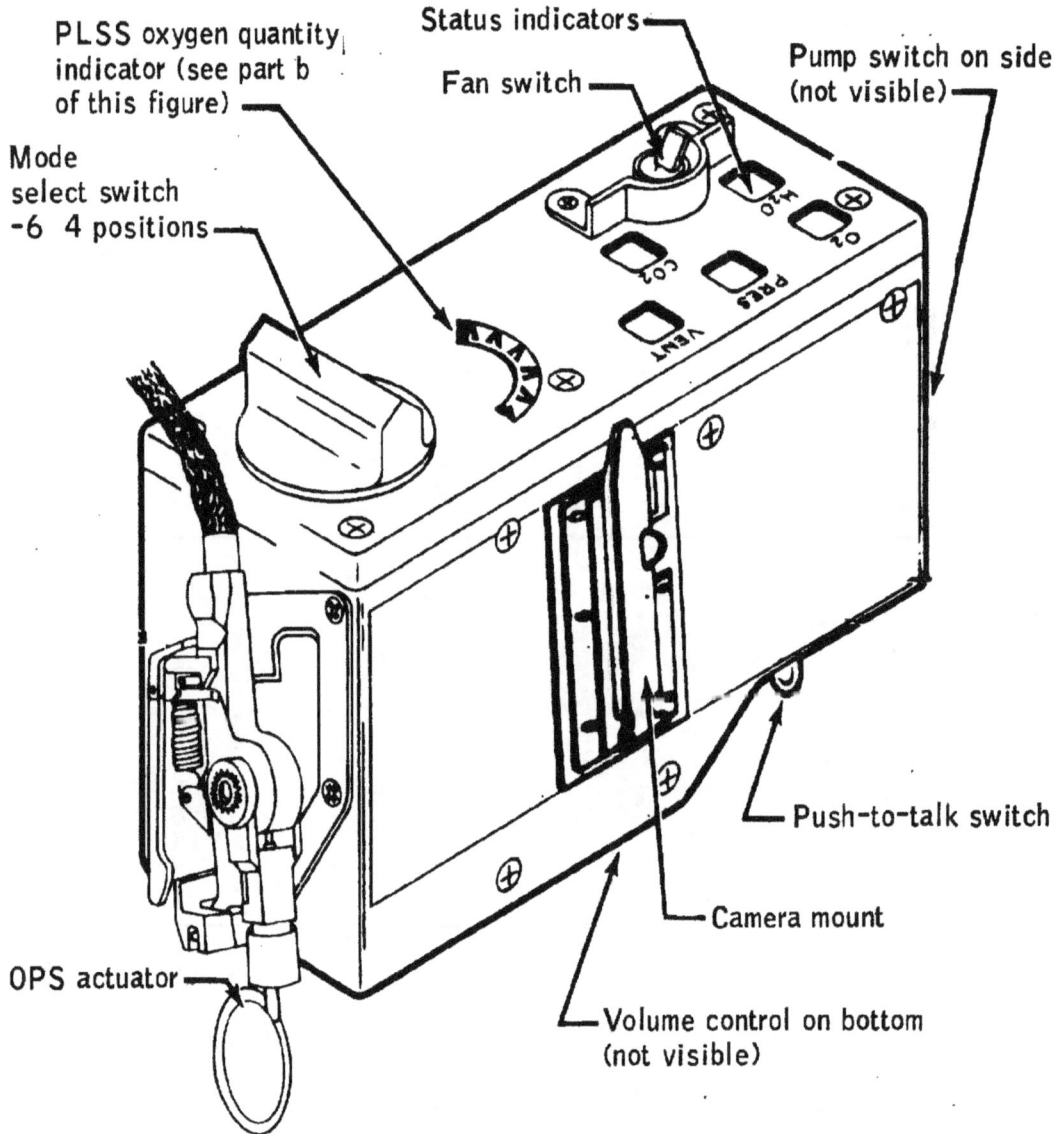

(a) Pictorial view of main elements.

Figure 2-43.- Remote control unit.

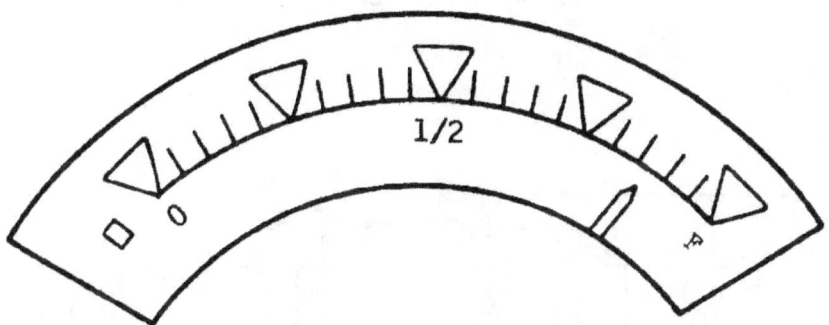

Each increment of indicator represents 68 psia.

Marking	Oxygen bottle pressure range, psia[a]
0	150 ± 68
1/4	490 ± 68
1/2	825 ± 68
3/4	1163 ± 68
F	1500 ± 68

[a] With RCU in a horizontal position and zero g.

(b) Oxygen quantity indicator markings and accuracies.

Figure 2-43.- Continued.

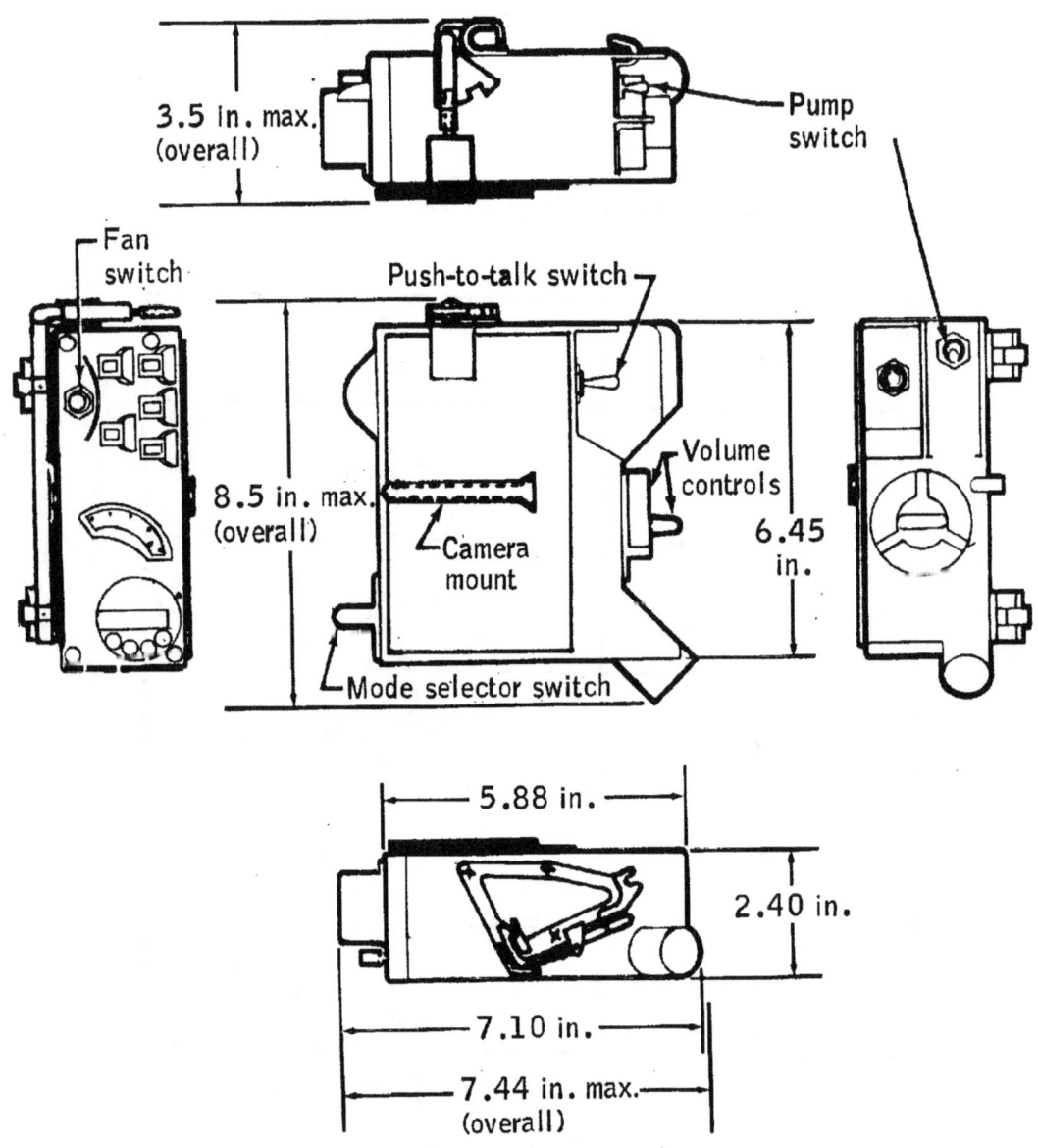

(c) Dimensions.

Figure 2-43.- Concluded.

to check his RCU and determine the problem area. Each warning symbol is a key to corrective action as follows.

Function	Indicator label	Symbol	Indicated action
High oxygen flow	O_2	O	Actuate OPS
Low PGA pressure	Pressure	O	Actuate OPS
Low vent flow	Vent	P	Purge
Low feedwater pressure	H_2O	A	Open auxiliary feedwater shut-off valve or use BSLSS as required

In addition to the above functions, the RCU provides a mounting point for the OPS actuator cable and the camera bracket.

2.6   OXYGEN PURGE SYSTEM

The OPS (fig. 2-44) supplies the EMU with oxygen purge flow and pressure control for certain failure modes of the PLSS or PGA during EVA. In the event of a PLSS failure, the OPS flow is regulated to 3.7 ± 0.3 psid for 30 minutes to provide breathing oxygen to the crewman, to prevent excessive carbon dioxide buildup, and to provide limited cooling. In this mode, the crewman sets his purge valve in the high-flow position (8.1 pounds per hour). In a second mode, the OPS may be used to provide make-up flow to the PLSS oxygen ventilating circuit via the PGA at flow rates of 0.07 to 2.0 pounds per hour. Finally, the OPS can be used in conjunction with the BSLSS (as described in section 2.7) to provide a 1.25-hour supply of purge flow for a crewman with a failed PLSS. For this mode, the crewman sets his purge valve in the low-flow position (4.0 lb of O_2 per hour).

In the lunar EVA configuration, the OPS is mounted on top of the PLSS (fig. 2-1). For normal EV activity from the command module, the OPS is worn in the helmet-mounted mode as shown in figure 2-45. During contingency EV transfer from the lunar module, however, the OPS is attached by straps to the lower front torso of the PGA (fig. 2-46).

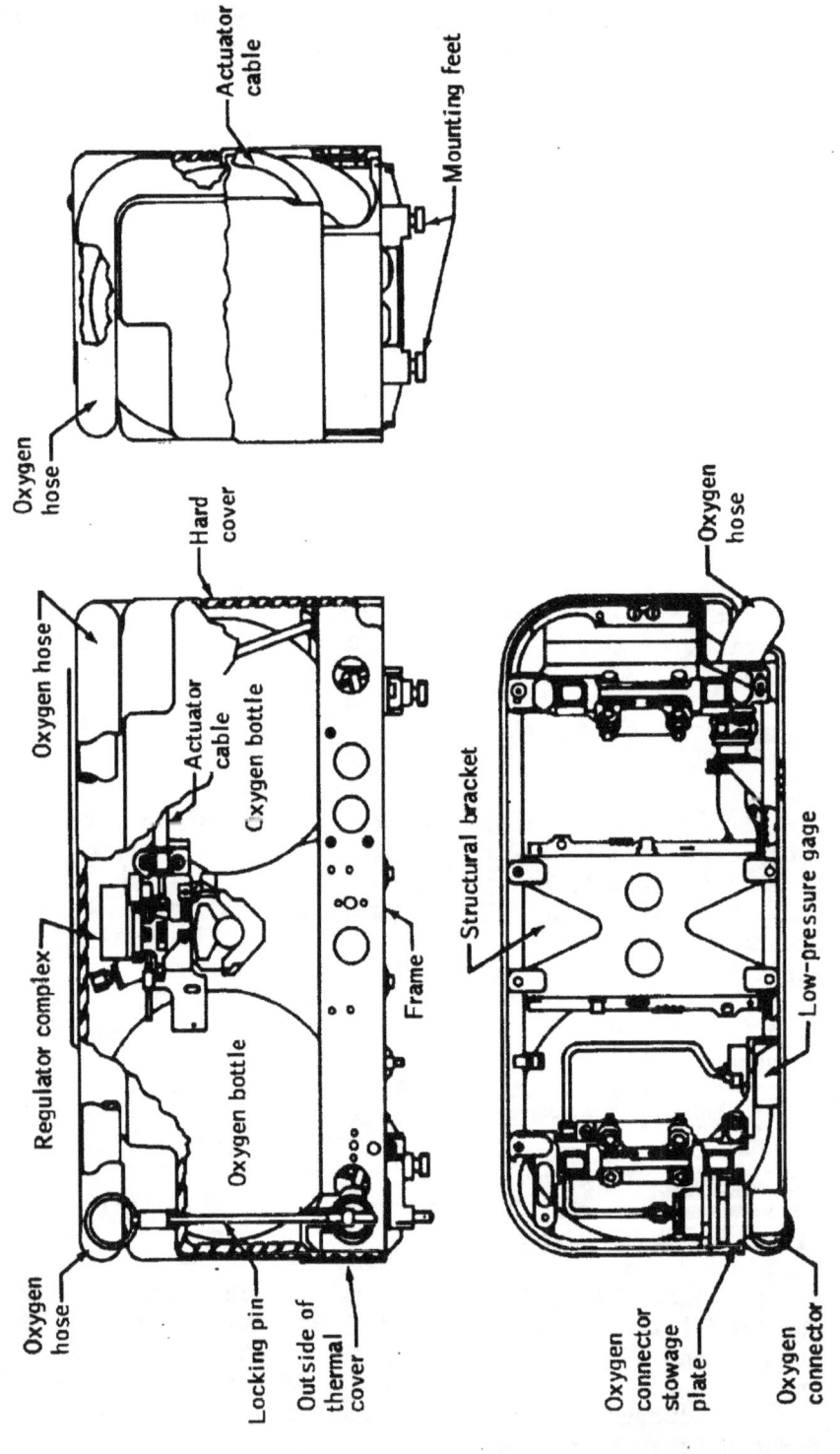

Figure 2-44.- Oxygen purge system, -3 configuration.

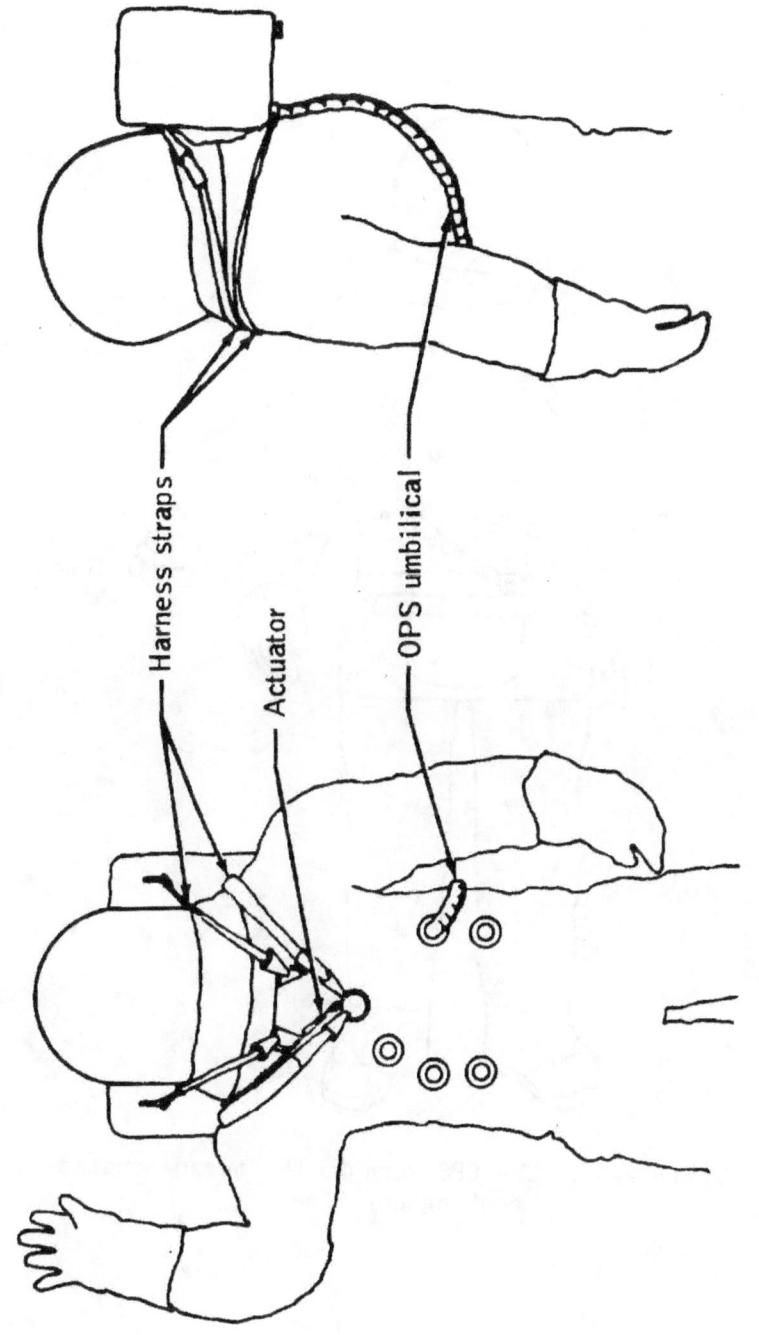

Figure 2-45.- The OPS worn in the helmet-mounted mode.

2-108 CSD-A-789-(1) REV V

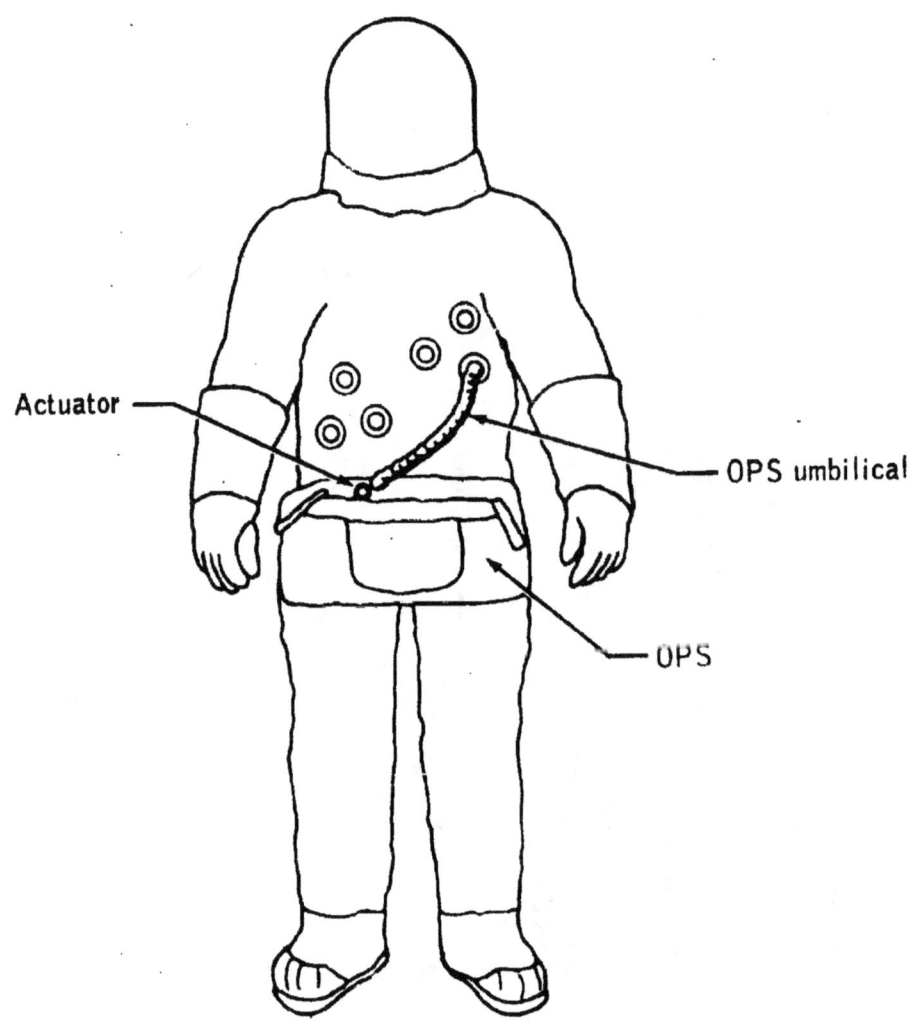

Figure 2-46.- The OPS worn in the torso-mounted contingency mode.

A schematic representation of the OPS is shown in figure 2-47. The OPS consists of two interconnected, spherical, high-pressure oxygen bottles (total of 5.1 pounds of usable oxygen at 5880 ± 80 psia and 70° F), a pressure regulator assembly, a fill fitting, a high-pressure gage, a delta-pressure gage, a suit connector and hose, a suit connector stowage plate, a shutoff valve, and an actuator cable and handle. The OPS has no communications capability, but provides the hard mount for the PLSS antenna. The OPS used for Apollo 15 and subsequent missions differs from the OPS used on Apollo 14 in that attachment points for the PLSS harnesses have been moved to permit helmet mounting. Also the oxygen outlet temperature control capability incorporated in the OPS for all missions through Apollo 13 has been deleted. Thus the heater, control circuitry, terminal board, temperature sensor, power switch, and battery have been removed.

The OPS is not rechargeable during a mission. The high-pressure gage is used to monitor bottle pressure during ground charge and during preoperational checkout. The delta-pressure gage is used during preoperational checkout to verify regulated flow through a 0.44- to 0.70-pound-per-hour orifice mounted on the connector stowage plate.

2.7 BUDDY SECONDARY LIFE SUPPORT SYSTEM

The BSLSS enables two EVA crewmen to share the water cooling provided by one PLSS following loss of cooling capability in the other PLSS. The system is shown schematically in figure 2-48 and in use by two crewmen in figure 2-49. The BSLSS is made up of six principal components.

a. Two water hoses 8-1/2 feet long and 3/8 inch inside diameter to carry the coolant flow between the good PLSS and the other crewman

b. A normal PLSS water connector on one end of the double hose

c. A flow-dividing connector on the other end of the double hose consisting of an ordinary PLSS water connector coupled with a receptacle to accept a PLSS water connector

d. A 4-1/2-foot restraint tether with hooks for attachment to the PGA LM restraint loops

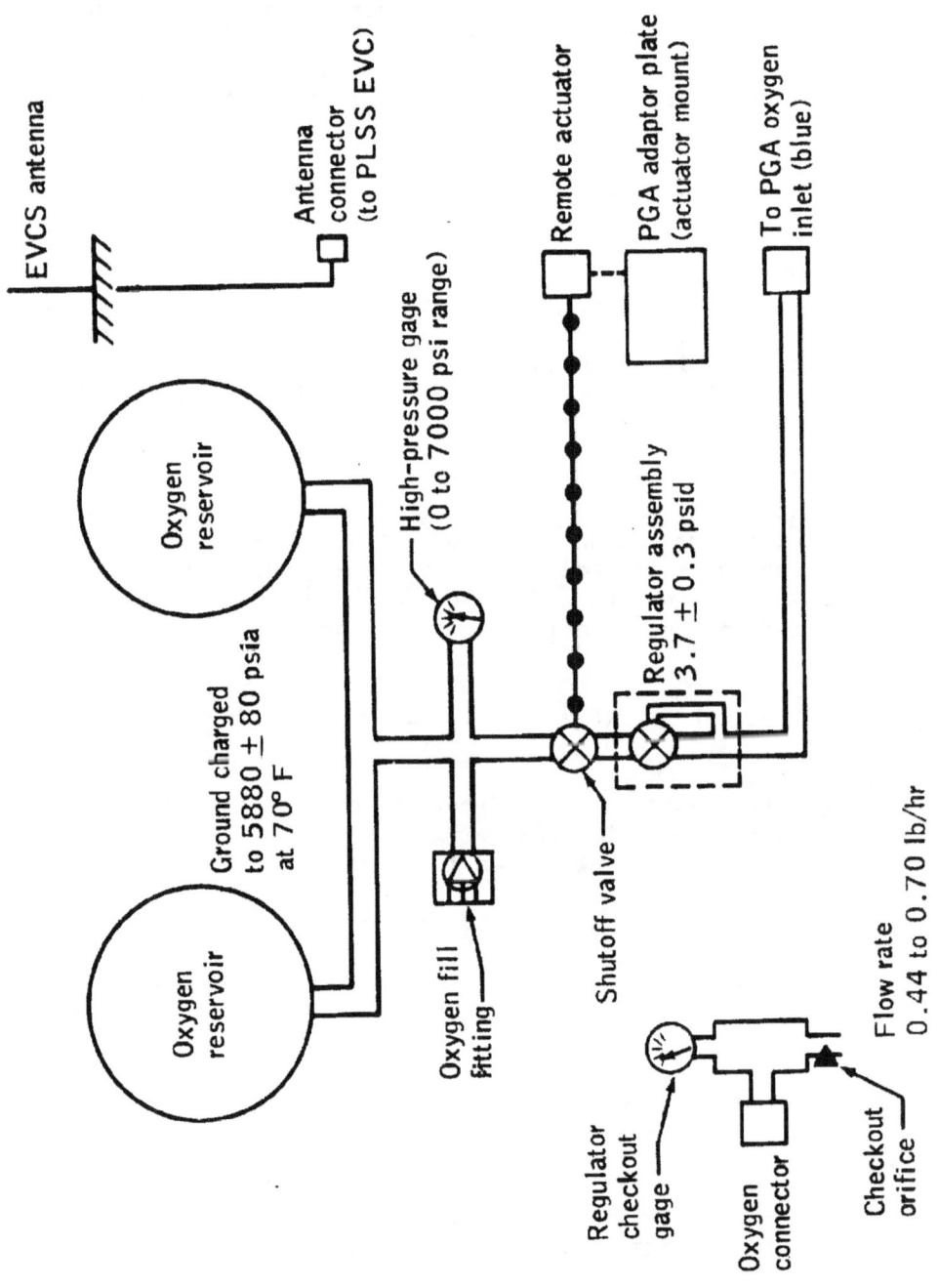

Figure 2-47.- Oxygen purge system schematic.

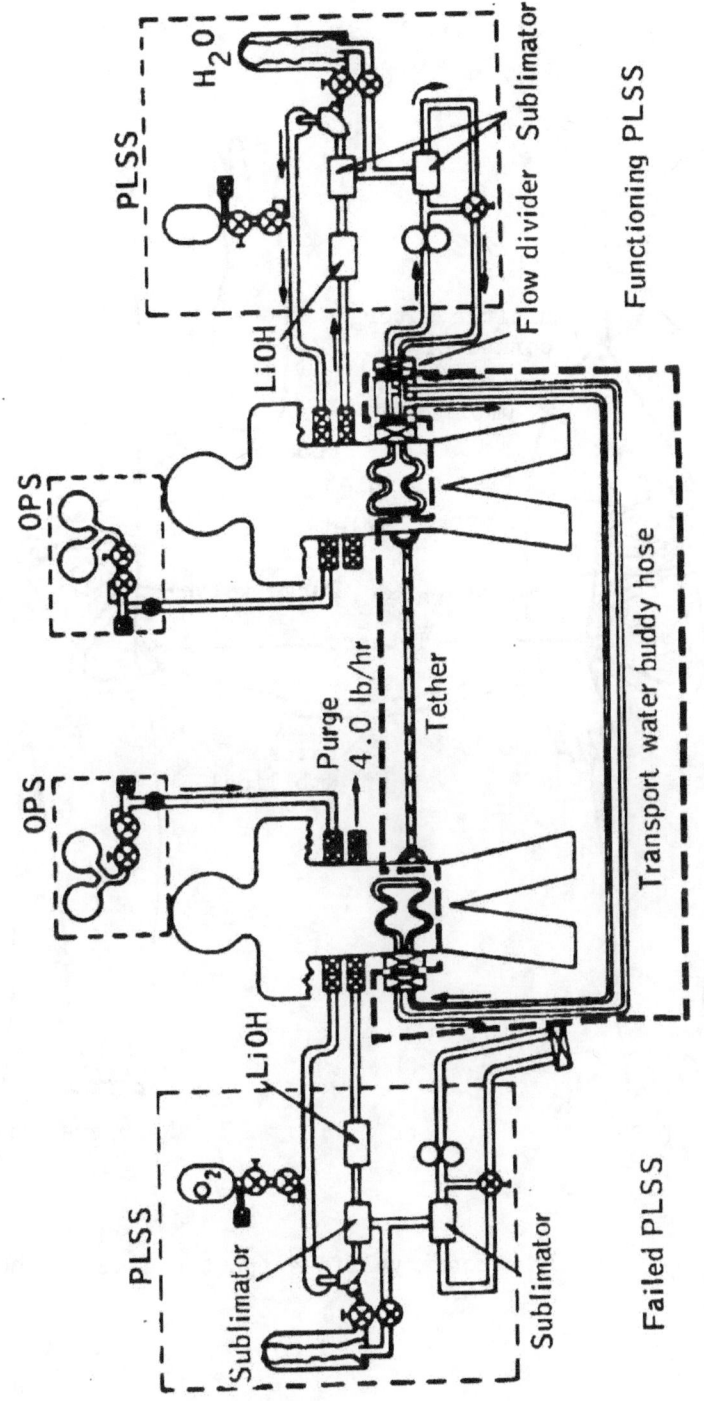

Figure 2-48.- Buddy secondary life support system schematic.

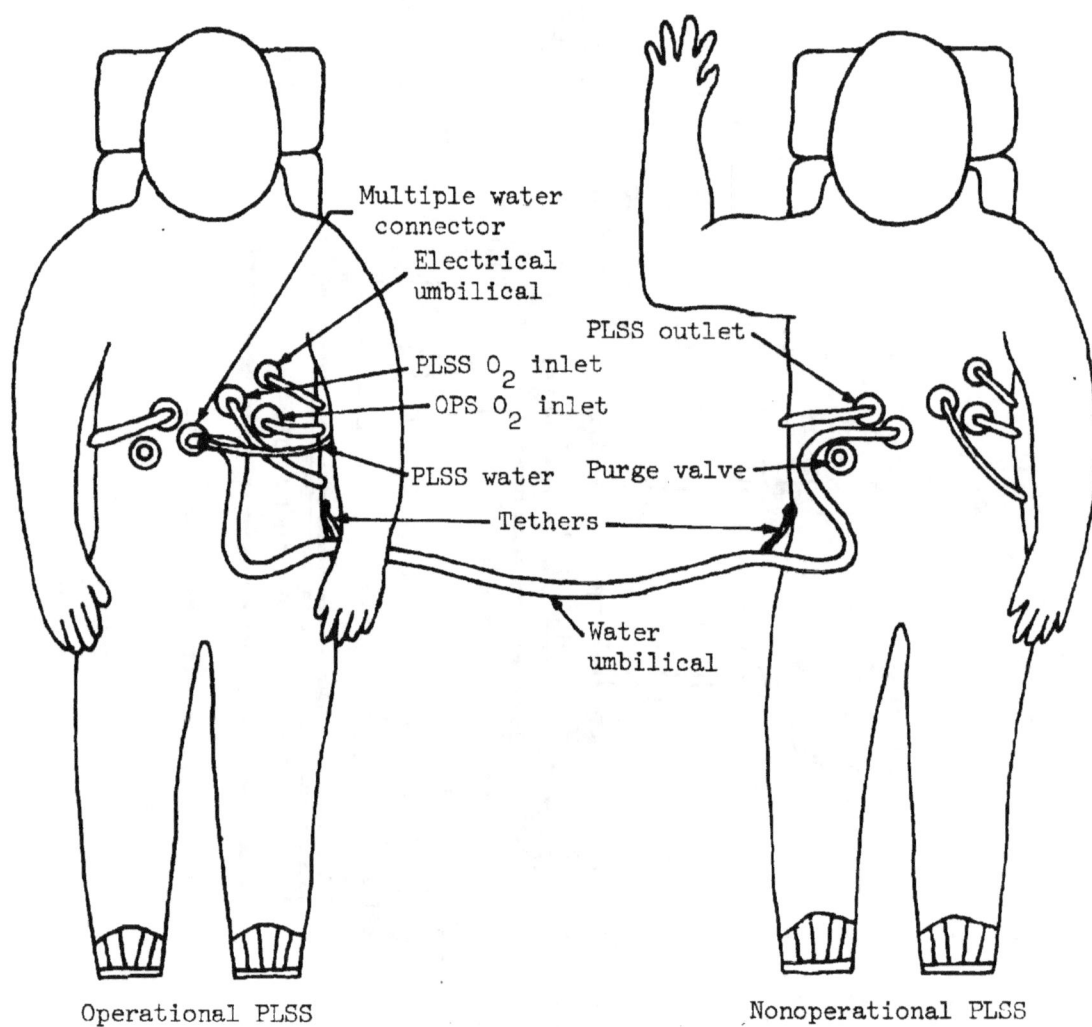

Figure 2-49   Buddy secondary life support system connected.

CSD-A-789-(1) REV V   2-11

    e. A thermal sheath the length of the hoses with tether breakouts 2 feet from each end

    f. A thermal pouch for stowage of the assembly on the PLSS during EVA and in the LM cabin during non-EVA periods (fig. 2-50)

The BSLSS hose stowage is illustrated in figure 2-50.

2.8   PRESSURE CONTROL VALVE

A pressure control valve (PCV) controls PGA pressure during normal EV transfer from the command module. This is a relief valve installed in one of the PGA outlet gas connectors prior to EVA. A purge valve is installed in the other outlet gas connector. Oxygen is supplied from the command module environmental control system at a flow rate of 10 to 12 pounds per hour via an umbilical to one of the PGA gas inlet connectors. The OPS, worn in the helmet-mounted configuration, provides a backup oxygen supply. The PCV contains a spring-loaded poppet which senses suit pressure and unseats, dumping a sufficient amount of suit oxygen to space to maintain suit pressure in the 3.5- to 4.0-psid range. The PCV is also sized to prevent suit pressure from falling below 3.2 psid in the event the poppet fails open. The PCV is shown in a schematic representation in figure 2-51.

2.9   PLSS FEEDWATER COLLECTION BAG

Deleted

Amendment 2
11/5/71

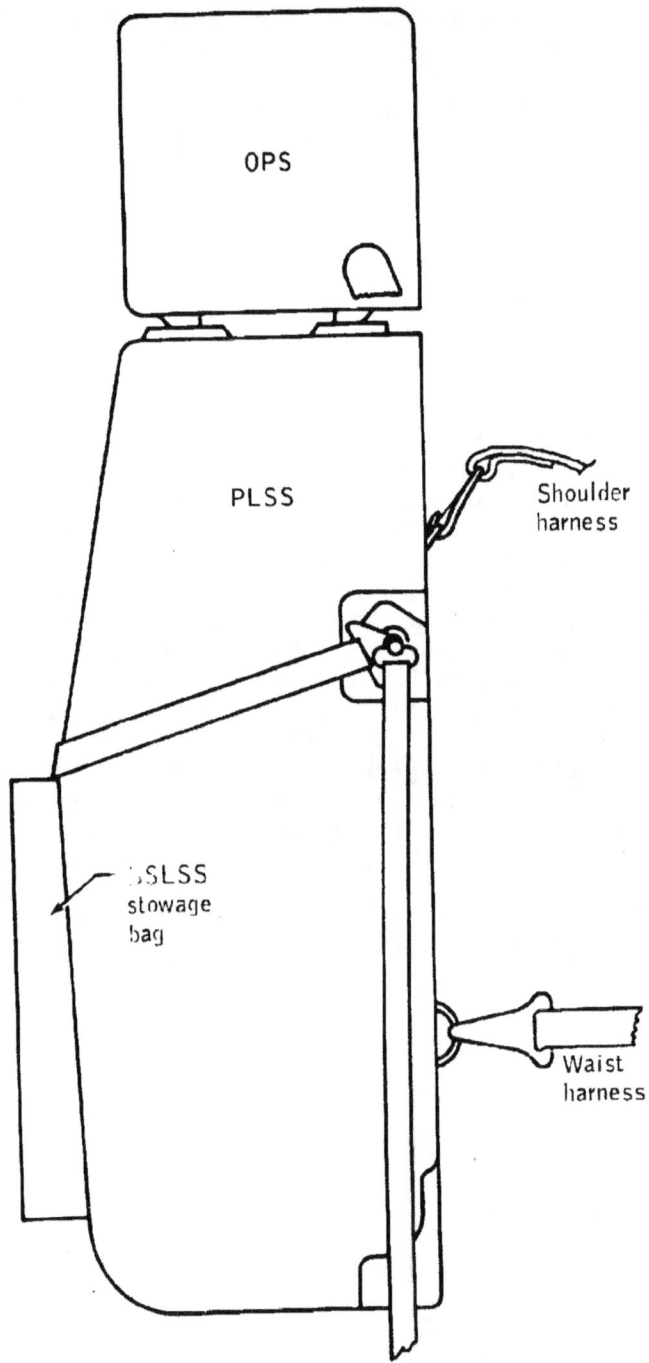

Figure 2-50.- BSLSS hose stowage.

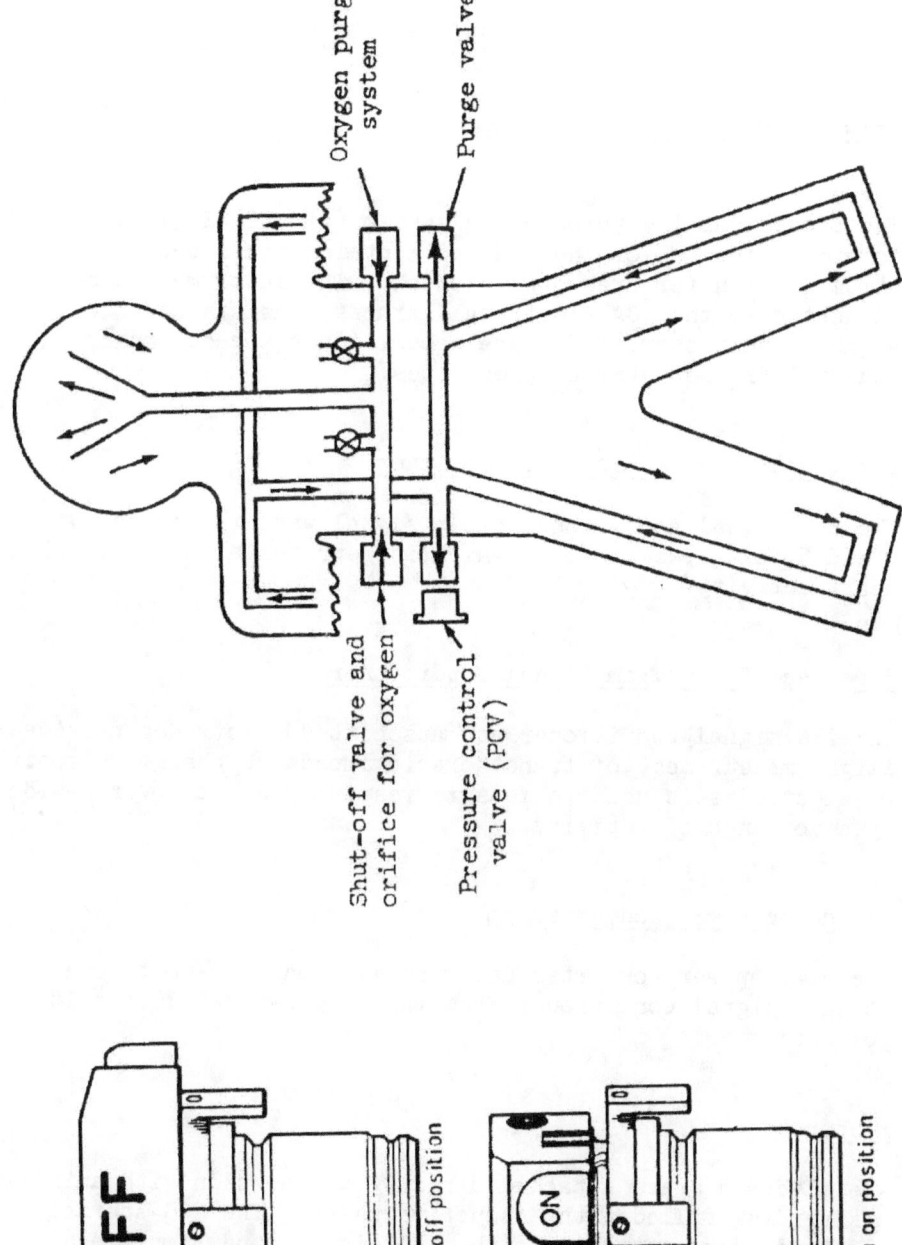

Figure 2-51.- Pressure control system.

2.10   BIOMEDICAL INSTRUMENTATION SYSTEM

The biomedical instrumentation system (fig. 2-52) is attached to either the CWG or the LCG and contains the necessary instrumentation for crew status check. The instrumentation connected to the PGA electrical harness consists of an ECG signal conditioner, ZPN signal conditioner, dc-dc converter, and axillary and sternal electrodes.

2.10.1   Electrocardiogram Signal Conditioner

The ECG signal conditioner has a signal wave ranging between 0 and 5 volts peak to peak which is representative of inflight heart activity.

2.10.2   Impedance Pneumogram Signal Conditioner

The ZPN signal conditioner and associated electrodes provide flight measurement of transthoracic impedance change. A pair of electrodes is used to measure respiration rate over a wide dynamic range of activity.

2.10.3   The dc-dc Power Converter

The dc-dc power converter delivers +10- and -10-volt power to each signal conditioner from the single-ended 16.8-volt power source.

2.10.4   Electrodes

The electrodes are attached directly to the skin with an adhesive disk filled with conductive paste. The ECG sternal electrodes are attached to the ECG signal conditioner and the ECG axillary electrodes are attached to the ZPN signal conditioner.

Amendment 2
11/5/71

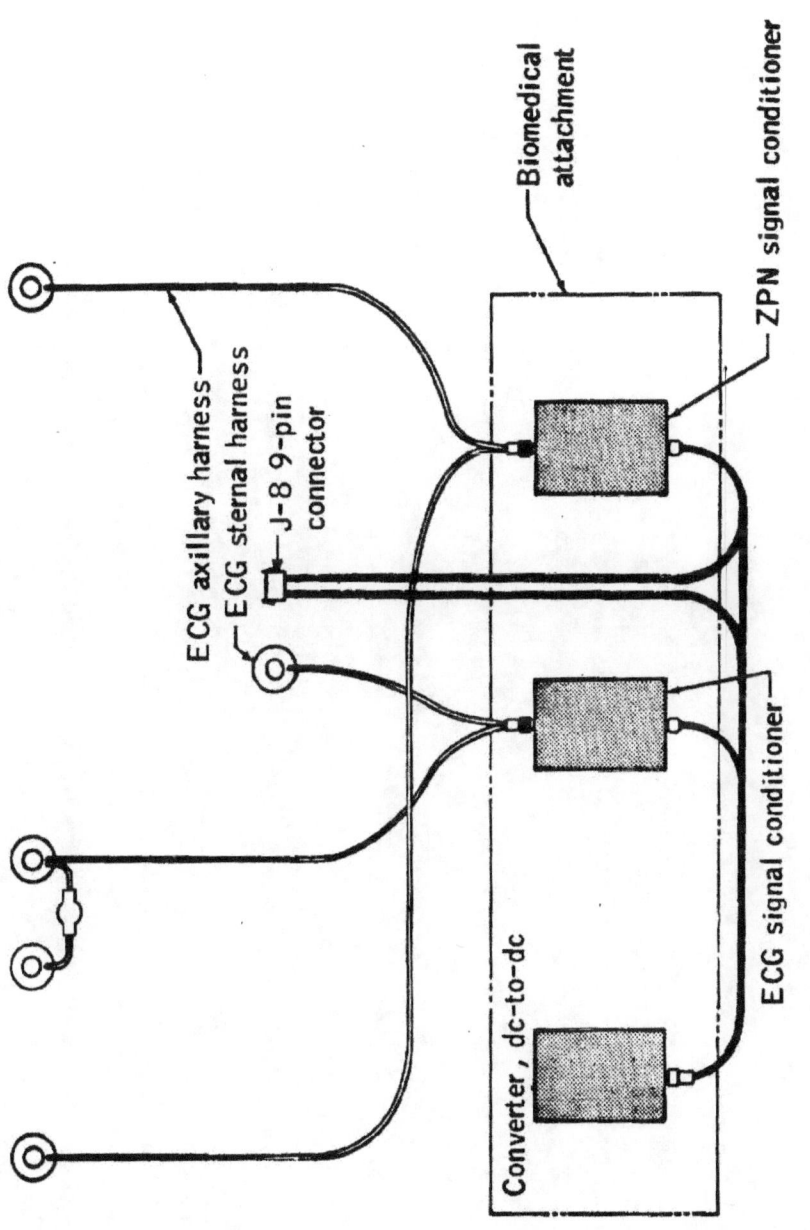

Figure 2-52.- Biomedical instrumentation system.

3.0 EXTRAVEHICULAR MOBILITY UNIT SYSTEMS

3.1 PRIMARY PRESSURIZATION AND VENTILATION

The EMU primary pressurization and ventilation system (fig. 3-1) is a closed-loop gas system which provides a habitable environment for the astronaut during Apollo EVA missions. A precharged oxygen bottle (1410 ± 30 psia) and pressure regulator pressurizes the system to 3.85 ± 0.15 psig and supplies the system with make-up oxygen to satisfy a 1200-Btu/hr metabolic load plus an EMU leakage factor for a 5-hour EVA design mission. The pressurized, breathable gas is forced through the loop at a rate of 6.00 cfm by a circulation pump. The circulated gas flows through the pressurizable porti of the PGA consisting of a TLSA, helmet, and a pair of gloves.

Within the pressurizable envelope, a ventilation distribution system directs the gas flow from the inlet connector to the helmet and the torso, if desired, down over the body to the limb extremities, then through noncrushable ducts to the outlet gas connector. The exhaust gas flows from the PGA to the PLSS through an umbilical.

Within the PLSS, the gas passes through a contamination-control assembly where odors are removed by activated charcoal. Carbon dioxide is removed by chemical reaction with LiOH, and foreign particles are filtered out by a peripheral Orlon filter. The oxygen passes from the contamination-control assembly to a sublimator which then cools the circulated oxygen. The cooled oxygen passes from the sublimator to the water separator where excess water entrained in the cooled oxygen is removed at a maximum rate of 0.508 lb/hr. The oxygen passes from the water separator to the fan/motor assembly for recirculation.

If a hypodermic injection is required, it is administered through the biomedical injection patch located on the left thigh. The patch is a self-sealing disk which prevents suit leakage as a result of the injection.

Suit pressure can be monitored continuously on a pressure gage installed on the left wrist of the PGA. The dial-indicating instrument is calibrated from 2.5 to 6.0 psid. In the event of suit overpressure, a pressure relief valve, located on the right thigh of the EV PGA and the left wrist

Amendment 2
11/5/71

To be determined

Figure 3-1.- EMU primary pressurization and ventilation system.

cone of the CMP PGA, opens at pressures of 5.00 to 5.75 psid and reseats at not less than 4.6 psid.

The flow of oxygen through the PLSS regulator assembly is limited to a maximum of 4.0 lb/hr at 1500 psia to protect the PGA against overpressurization in the event of a failed-open regulator. This is accomplished by an orifice between the regulator and the prime oxygen bottle and fill connector. The fill connector is a leak-proof, self-sealing, quick-disconnect connector used for recharging the primary oxygen subsystem. Recharge time from a 1425-psia source at 0° to 60° F is a nominal 75 minutes. An oxygen flow sensor gives an audible tone when PLSS primary oxygen flow exceeds a 0.50 to 0.65 lb/hr band and will remain actuated until the flow decreases to 0.50 to 0.65 lb/hr (a continuous high flow of 0.50 to 0.65 for 5 seconds is needed for actuation). A primary oxygen pressure transducer provides electrical signals to the oxygen quantity indicator for crew visual read-out and to the telemetry system of the PLSS.

Two additional pressure transducers are incorporated in the primary oxygen subsystem to monitor the PGA pressure. One transducer is used for telemetry monitoring, and the other activates an audible warning tone when PGA pressure drops below 3.10 to 3.40 psid.

3.2 LIQUID COOLING SYSTEM

The EMU oxygen pressurization and ventilation system removes body heat by carrying evaporated body perspiration from the PGA. To reduce body fluid loss and increase body cooling efficiency, the liquid cooling system is employed for transporting metabolic heat from the PGA. The liquid (water) cooling system (fig. 3-2) is a closed-loop system fed by a pressurized water reservoir. The reservoir is pressurized by the EMU pressurization and ventilation system, and a pump circulates the water through the closed-loop system at a nominal rate of 4.0 lb/min.

The water supplied by the PLSS passes through the inlet passage of the multiple water connector and circulates through the manifold and a network of polyvinylchloride tubing contained in the LCG. During the circulation process, the heat within the PGA is transferred by conduction to the water which returns through the outlet passage of the multiple

Amendment 2
11/5/71

To be determined

Figure 3-2.- EMU liquid cooling system.

water connector to the PLSS for cooling. The water within the PLSS is circulated through the sublimator to provide the cooling. The sublimator is supplied with expendable feedwater from the feedwater reservoir.

The feedwater is enclosed by a collapsible bladder within the reservoir with the exterior of the bladder exposed to the ventilation loop pressure through the water separator. This pressure provides the force required to supply feedwater to the sublimator. It also enables the portion of the feedwater reservoir external to the bladder to be used for the storage of waste water removed from the ventilation loop.

MSC-01372-2

NATIONAL AERONAUTICS AND SPACE ADMINISTRATION

APOLLO OPERATIONS HANDBOOK EXTRAVEHICULAR MOBILITY UNIT

JUNE 1971

VOLUME II
OPERATIONAL PROCEDURES
CSD-A-789-(2)
APOLLO 15-17

CREW SYSTEMS DIVISION
ORIGINAL ISSUE MAY 1969
REVISION III

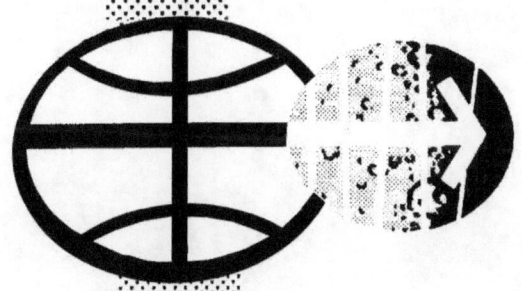

MANNED SPACECRAFT CENTER
HOUSTON, TEXAS

Manned Spacecraft Center
Crew Systems Division

MSC-01372-2

PROJECT DOCUMENT COVER SHEET

APOLLO OPERATIONS HANDBOOK

EXTRAVEHICULAR MOBILITY UNIT

VOLUME II — OPERATIONAL PROCEDURES

CSD	DATE
CSD-A-789-(2) III	May 1969

PREPARED BY:	James L. Gibson, Apollo Support Branch 5/14/69
APPROVED:	(BRANCH) Charles C. Lutz, Chief 5/16/69
APPROVED:	(DIVISION) Robert E. Smylie, Chief, CSD
APPROVED:	(CREW PROCEDURES DIVISION, EVA/IVA BRANCH, FCSD)

NO. OF PAGES 136

REVISIONS					
DATE	PREPARED BY	APPROVALS			REV. LETTER
		BRANCH	DIVISION	PROGRAM OFFICE	
Oct. 69	J. L. Gibson				Rev. I
Sept 70	J. L. Gibson 10/12				Rev. II
June 71	J. L. Gibson 7/1				Rev. III

CSD CSD-A-789-(2) III

MSC FORM 151 (REV MAY 67)

PROJECT DOCUMENT CHANGE/REVISION LOG
FOR CSD ORIGINATED DOCUMENT
NUMBER CSD-A-789(2) III

CHG. NO. / DATE	AUTHORITY FOR CHANGE	PAGES AFFECTED	BRIEF DESCRIPTION OF CHANGE	END ITEM/ SERIAL NUMBER AFFECTED
Rev. III	EMU 77A through 88A	All	Reorganized and rewritten to accommodate A7LB suit configuration and the -7 PLSS configuration	Apollo 15

ALTERED PAGES MUST BE TYPED & DISTRIBUTED FOR INSERTION

CSD-A-789-(2) III
APOLLO OPERATIONS HANDBOOK - EMU

CONTENTS

Section		Page
4.	OPERATIONAL PROCEDURES	4-1
4.1	PGA DONNING AND DOFFING	4-2
4.1.1	PGA Donning Procedures Before IV Use	4-2
4.1.2	PGA Doffing Procedures After IV Use	4-15
4.1.3	PGA Donning Procedures Before EV Use	4-20
4.1.4	PGA Doffing Procedures After EV Use	4-25
4.2	OPS OPERATIONS	4-29
4.2.1	OPS Checkout	4-29
4.2.2	OPS Donning Prior to Contingency Transfer	4-29
4.2.3	OPS Donning Before CMP EVA	4-30
4.2.4	OPS Activation	4-32
4.2.5	OPS Deactivation	4-32
4.2.6	OPS Doffing After CMP EVA	4-32
4.3	PLSS/OPS DONNING AND DOFFING	4-33
4.3.1	PLSS/OPS Donning	4-33
4.3.2	PLSS/OPS Doffing in Pressurized Environment	4-36
4.3.3	PLSS/OPS Doffing in Unpressurized Environment	4-40
4.4	PLSS COMMUNICATIONS CHECK	4-45
4.5	PRESSURE INTEGRITY CHECK	4-47
4.6	PLSS NORMAL OPERATING MODES	4-49
4.6.1	PLSS Activation	4-49
4.6.2	Wet Sublimator Restart	4-50
4.6.3	PLSS Deactivation	4-50
4.7	PLSS RECHARGE AND FEEDWATER REMOVAL PROCEDURES	4-52
4.7.1	Battery Replacement	4-52

BASIC DATE May 1969   **CHANGE DATE** June 1971   **PAGE** iii

APOLLO OPERATIONS HANDBOOK - EMU

Section		Page
4.7.2	Removal of LiOH Cartridge From Stowage Container	4-53
4.7.3	LiOH Cartridge Change	4-54
4.8	PURGE VALVE OPERATION	4-56
4.8.1	Activation Procedures for Purge Valve	4-56
4.8.2	Flow Adjustment Procedures	4-56
4.8.3	Shutoff Procedures	4-56
4.8.4	Purge Valve Removal Procedures	4-57
4.9	PRESSURE CONTROL VALVE OPERATION	4-58
4.9.1	PCV Installation	4-58
4.9.2	PCV Activation	4-58
4.9.3	PCV Deactivation	4-58
4.9.4	PCV Removal Procedures	4-58
4.10	EMU INFLIGHT MAINTENANCE	4-59
4.10.1	PGA and LEVA Inflight Examination and Maintenance	4-59
4.10.2	Bioinstrumentation Inflight Repair	4-64
4.11	MISCELLANEOUS PROCEDURES	4-66
4.11.1	PLSS Gas Trap Activation	4-66
4.11.2	Camera Mounting on RCU	4-67
4.11.3	BSLSS Bag to PLSS Donning	4-68
4.11.4	BSLSS Bag to PLSS Doffing	4-68
4.11.5	BSLSS Stowage on LRV	4-68
4.11.6	BSLSS Donning and Activation Procedure	4-69
4.11.7	BSLSS Disconnect Procedure	4-69
4.11.8	Donning and Doffing Helmet Shield	4-70
4.11.9	Donning and Doffing Neck Dam	4-70
4.11.10	Insuit Drinking Device (ISDD) Installation and Use	4-71

BASIC DATE May 1969   **CHANGE DATE** June 1971

CSD-A-789-(2) III
APOLLO OPERATIONS HANDBOOK - EMU

Section		Page
4.11.11	OPS Oxygen Usage for Metabolic Makeup — LM and CM	4-74
4.11.12	OPS Bleed Down — LM and CM	4-74
5.	<u>EMU MALFUNCTION PROCEDURES FOR LUNAR SURFACE EVA</u>	5-1
5.1	INTRODUCTION	5-1
5.2	FORMAT OF MALFUNCTION PROCEDURES	5-1
5.3	MALFUNCTION SYMPTOMS FOR EMU USING PLSS SV 706100-7	5-3

ABBREVIATIONS

acfm	actual cubic feet per minute
AOH	Apollo Operations Handbook
BSLSS	Buddy Secondary Life Support System
CDR	Commander
cfm	cubic feet per minute
CM	command module
CMP	command module pilot
comm	communications
CSM	command/service module
CWG	constant wear garment
"DES" H_2O	descent water
ECG	electrocardiogram
ECS	environmental control system
EMU	extravehicular mobility unit
EV	extravehicular
ISDD	insuit drinking device
LEVA	lunar extravehicular visor assembly
FCS	fecal containment system
ITMG	integrated thermal micrometeoroid garment
IV	intravehicular
IVA	intravehicular activity
kHz	kilohertz (thousand cycles per second)
LCG	liquid cooling garment
LiOH	lithium hydroxide
LM	lunar module
MSFN	Manned Space Flight Network

OPS	oxygen purge system
PGA	pressure garment assembly
PHA	pressure helmet assembly
PLSS	portable life support system
pos.	position
POS	primary oxygen system of PLSS
press.	pressure
psia	pounds per square inch absolute
psid	pounds per square inch differential
psig	pounds per square inch gage
PTT	push to talk
RCU	remote control unit
rcvr	receiver
SC	spacecraft
sel.	selector
SSC	space suit communication
SW.	switch
TLSA	torso limb suit assembly
TM	telemetry
UCTA	urine collection and transfer assembly
UTS	urine transfer system
vlv	valve
VOX	voice operated transmitter
WMS	waste management system
xducer	transducer
xmtr	transmitter
ZPN	impedance pneumograph

ΔT   change in temperature

ΔP   change in pressure

CSD-A-789-(2) III
APOLLO OPERATIONS HANDBOOK - EMU

FOREWORD

This handbook, sections 4 and 5, of the Apollo Operations Handbook (AOH) series, is bound separately as Volume II and pertains only to the Extravehicular Mobility Unit (EMU). Volume I of the AOH contains sections 1, 2, and 3, and pertains to the description of the EMU.

The operational procedures are written specifically for Apollo J-mission EMU hardware and will be superseded by a mission-oriented checklist which is reviewed by Crew Systems Division (CSD) for unique hardware procedures. Both the operational procedures and the malfunction procedures will be updated for the mission.

This handbook is composed of two major sections:

a. Section 4 provides a step-by-step operational procedure for activation and deactivation of EMU subsystems.

b. Section 5 provides emergency procedures for critical EMU symptoms and continuing diagnosis during EVA and post-EVA to determine the EMU malfunction (to be supplied).

Insuiries concerning this handbook should be addressed to Crew Systems Division, Systems Engineering Branch, MSC, mail code EC2 or to Crew Procedures Division, EVA/IVA Procedures Branch, mail code CG3.

SECTION 4

OPERATIONAL PROCEDURES

This section includes the procedures for activation and deactivation of the EMU equipment. Techniques in donning may vary within an individual step, but the sequence of the steps outlined should not be changed.

CSD-A-789-(2) III
APOLLO OPERATIONS HANDBOOK - EMU

STEP	PROCEDURE	REMARKS
4.1	PGA DONNING AND DOFFING	The comfort gloves and wristlets are provided as crew preference items and may be donned and doffed as convenient to the wearer.
4.1.1	PGA Donning Procedures Before IV Use	If the bioinstrumentation system is required, shave electrode sites where necessary and shower after shaving. Placement of biomedical sensors or electrodes shall be done under medical direction (see note).
4.1.1.1	Donning FCS	
a.	Don FCS as conventional underwear, and place genitals through front opening.	
b.	Position FCS around waist and thighs to insure a seal. Ensure a snug fit with no sag in the crotch portion of the FCS.	
4.1.1.2	Donning Bioinstrumentation and CWG	The steps that prescribe the installation of the bioinstrumentation may be skipped when their use is not required (Steps d., e., and h. through k; see note).
a.	Open front buttons of CWG.	
b.	Don CWG to waist as conventional long underwear.	NOTE: Initial placement of biomedical sensors and electrodes is a pre-flight procedure.
c.	Adjust feet into integrated socks.	
d.	Clean shaved electrode sites with alcohol and let dry.	

4.1 PGA DONNING AND DOFFING

CSD-A-789-(2) III
APOLLO OPERATIONS HANDBOOK - EMU

STEP	PROCEDURE	REMARKS
e.	Assemble signal conditioners, biomedical harness with biomedical belt, and test as a system.	Use caution in mating biomedical connector to signal conditioners. Connectors are keyed for proper fit.
f.	Insert bioinstrumentation leads through slots on chest area of CWG. Don top half of CWG, and insert harness connectors (blue and yellow) through holes in CWG. Snap biomedical belt to CWG.	When transferring bioinstrumentation between CWG and LCG, or when reinstalling after removal, the signal conditioners should not be removed from the pockets in the biomedical belt. Only the nine-pin Airlock connector, sternal electrode harness (blue code), and the auxiliary harness (yellow code) should be disconnected. The color code of the harness-to-signal conditioner should be observed when reinstalling the biomedical belt. The electrodes are not removed from skin to change garments.
g.	Button front opening.	
h.	Connect electrode harness to proper color-coded signal conditioner. Dot on electrode harness is facing out.	
i.	Perform functional check.	
j.	Adjust ECG gain from normal signal to 40 percent of full scale.	
k.	Adjust ZPN gain to allow maximum inhalation to be at or near full scale.	

4.1 PGA DONNING AND DOFFING

BASIC DATE May 1969 **CHANGE DATE** June 1971 **PAGE** 4-3

CSD-A-789-(2) III
APOLLO OPERATIONS HANDBOOK-EMU

STEP	PROCEDURE	REMARKS
4.1.1.3	Donning UCTA	
a.	Obtain proper size cuff and roll down to cuff flange after removing flange from UCTA.	
b.	Place penis in flange and roll cuff on.	
c.	Attach cuff half of the UCTA flange to the UCTA by depressing the flange release lever and twisting the two halves together.	
d.	Aline the colored waistband patches with the corresponding colors on the UCTA.	Place waistband of harness across small of back.
e.	Attach the crotch strap to the UCTA.	Strap length can be adjusted at mating point. Verify that all straps are attached.
f.	Connect hand pump to drain hose and evacuate all gas from the UCTA through the drain hose.	This step is required for donning at 14.7-psia ambient pressure only; not required for 5-psia donning.
4.1.1.4	Donning TISA	
4.1.1.4.1	<u>Donning EV A7LB TISA.-</u>	See paragraph 4.1.1.4.2 for CMP A7LB TISA donning instructions.
a.	Remove protective covers, gas connector caps, etc., from TISA as required, and stow.	
b.	Open ITMG slide fastener (zipper) cover flap.	

4.1 PGA DONNING AND DOFFING

BASIC DATE May 1969   CHANGE DATE June 1971   PAGE 4-4

CSD-A-789-(2) III
APOLLO OPERATIONS HANDBOOK - EMU

STEP	PROCEDURE	REMARKS
c.	Fully open both restraint zippers and the pressure sealing zipper closure.	Ascertain that the sliders and lanyards are under the shoulder restraint cable on the upper right chest area.
d.	Attach the donning-assist-hook-lanyard Velcro to the Velcro on the ITMG entrance closure flap.	
e.	Grasp TLSA, carefully place left foot into TLSA, and manipulate foot to enable easy insertion into boot.	
	NOTE	
	Use the donning strap located at the back and at the top of each ITMG boot as necessary to aid boot donning operations.	
f.	Repeat step e for right foot.	
g.	Pull upwards on suit until the upper part of the lower half of the suit is just below the fold immediately below the buttocks.	
h.	Connect the UCTA in-line connector.	Roll abrasion cover over the engaged urine connector.
	CAUTION	
	The connectors must be properly alined before engagement, or damage to the pins and seal will result.	
i.	Connect biomedical harness electrical branch to suit electrical harness.	

4.1 PGA DONNING AND DOFFING

BASIC DATE May 1969   CHANGE DATE June 1971   PAGE 4-5

CSD-A-789-(2) III
APOLLO OPERATIONS HANDBOOK - EMU

STEP	PROCEDURE	REMARKS
j.	Remove water connector plug.	
k.	Engage the LCG water connector to the TLSA multiple water connect receptacle, if LCG is worn.	Disregard this step if LCG is not worn. If water connector proves difficult to engage, check to make sure the locking balls are retracted and the locking ring on the receptacle is in the open position. Verify that the LCG connector is fully engaged and the receptacle fully connected.
l.	Make certain that PGA locking ring receptacle is in the CLOSE position.	
m.	Insert hands into shoulder convolutes, and lower and position head into neck ring while simultaneously slipping hands fully into arms and through wrist disconnects.	This and the following steps have proven satisfactory for the majority of test subjects. Alternate methods, such as full insertion of head through neck ring, or insertion of one arm and then the combined insertion of second arm and head, are acceptable provided the load on the TLSA, and/or the slide fastener assemblies is not increased.
n.	Assume a semierect position and slide the upper torso down over the shoulder and back.	
o.	Gradually assume an erect standing position, and, at the same time, work the lower half of the suit up over the buttocks. Keep pushing the CWG or LCG and UCTA down inside the TLSA.	

4.1 PGA DONNING AND DOFFING

BASIC DATE May 1969   CHANGE DATE June 1971   PAGE 4-6

CSD-A-789-(2) III
APOLLO OPERATIONS HANDBOOK - EMU

STEP	PROCEDURE	REMARKS
p.	After the bottom half of the TLSA is donned over the buttocks, reach back and remove the donning assist-hook lanyard from the Velcro attachment point. Pull the upper half of the TLSA around the body to the left in a counterclockwise position, and engage the assist hook over the front end of the right LM tether bracket. This will enable the subject to move his body within the suit and position the suit relative to his body.	
q.	Adjust the TLSA over the body and tuck in the excess material of the CWG or LCG and UCTA. Then, reach around the body and pull the lower half of the pressure-sealing closure up and the top half of the pressure-sealing closure down.	
r.	Close the pressure-sealing zipper by moving the pressure-sealing slider down and around the back while leaning in the direction of the slider direction. Pull the slider around the left side of the TLSA and up against the pressure-sealing closure lock, and engage and secure zipper lock. WARNING Check to ensure zipper lock is fully engaged by pulling out on its lanyard and attempting to disengage the assembly.	Make sure that the CWG or LCG and UCTA are tucked in and out of the way of the slider. While closing the pressure-sealing zippers, grasp ahead of the red lanyard on the pressure-sealing slider and pull the slider down until the lanyard can be reached (approximately 10 inches). Gradually move the slider around the closure keeping the hand as close to the slider as possible. Never force the slider. If it seems to stick, back the slider off slightly, check to see if there is any CWG, LCG, or UCTA material in the way, and then pull again. The slider will stick if the lips of the closure are folded in. Backing off the slider slightly and leaning in the direction of the slider will generally clear this condition and the slider will close more easily.

4.1 PGA DONNING AND DOFFING

CSD-A-789-(2) III
APOLLO OPERATIONS HANDBOOK - EMU

STEP	PROCEDURE	REMARKS
		The slider lock will not operate until the slider is fully seated against the stop.
		When the zipper is fully closed, the slide yoke is hinged over the lock and the zipper latch is moved into its lock position by the cloth lanyards.
s.	Close vertical restraint zipper by grasping the neck ring and exerting an upward force while simultaneously pulling downward on the blue lanyard until the slider reaches its stop at the right waist.	Engage donning-aid hooks to assist in the alinement of the vertical restraint zipper. While closing the restraint zippers, constantly tuck in the bladder material with two or three fingers ahead of the slider as it is moved to prevent binding.
t.	Disconnect the donning lanyard (yellow) hook from the LM tether hook and reconnect it to its Velcro attachment point.	
u.	Grasp the blue lanyard on the waist restraint zipper and pull the slider horizontally around to its stop at the right waist. Relieve the load on the slider to ease fastening by constantly leaning in a direction that would be slightly ahead of the slider as it travels.	

4.1 PGA DONNING AND DOFFING

CSD-A-789-(2) III
APOLLO OPERATIONS HANDBOOK - EMU

STEP	PROCEDURE	REMARKS
v.	Engage and seat restraint zipper slider into the slot in the restraint zipper lock and secure lock.	Lock the slider in place by placing the thumb upon the top of the lock assembly and the index finger over the red striker and squeezing until the lock-lock tab snaps into the lock position.
	WARNING	
w.	Verify that restraint zipper lock assembly is fully engaged and locked.	
	CAUTION	
	Close ITMG entrance closure over zipper and engage hook-and-pile fastener tape strips.	
	To engage the electrical connectors, aline the connector halves and gently slide the halves together until the locks engage. Under no circumstances shall electrical connectors be forcibly engaged.	
x.	Don communications carrier, and unsnap electrical lead retainer from liner and connect electrical lead to upper branch of electrical harness. Then tuck connector and lead to comfortable position within TLSA, and connect electrical umbilical to TLSA and test system.	Do not bend mike boom more than 80° from center. Bend mike boom from the middle section only. Do not straighten or bend the mike boom unless necessary.

4.1 PGA DONNING AND DOFFING

BASIC DATE May 1969   CHANGE DATE June 1971   PAGE 4-9

CSD-A-789-(2) III
APOLLO OPERATIONS HANDBOOK - EMU

STEP	PROCEDURE	REMARKS
	CAUTION	
	Gas umbilicals must be inserted straight into the gas connectors to prevent excessive side loading and subsequent damage to gas connector O-ring seals. To prevent an inadvertent overpressure, always connect the outlet umbilical first.	
y.	Connect outlet gas umbilical to exhaust gas connector.	
z.	Connect gas inlet umbilical to inlet gas connector.	
aa.	Provide ventilation flow.	
4.1.1.4.2	Donning CMP A7LB TLSA.-	
a.	Remove protective cover from TLSA electrical connector and two protective plugs from gas connectors and stow.	
b.	Loosen torso tiedown strap.	
c.	Fully open both the restraint and pressure-sealing zippers.	Remove donning lanyards from UCTA/biomedical injection access flap. If unassisted, insert the red suit-donning lanyard through the pull tab of restraint zipper slider and the blue suit-donning lanyard through the pull tab of pressure-sealing zipper slider.
d.	Grasp TLSA at sides. Do not pick up TLSA at pressure-sealing zipper location. Place both feet into PGA and work feet into the boots.	

4.1 PGA DONNING AND DOFFING

BASIC DATE May 1969   CHANGE DATE June 1971   PAGE 4-10

CSD-A-789-(2) III
APOLLO OPERATIONS HANDBOOK - EMU

STEP	PROCEDURE	REMARKS
e.	Pull excess liner material up around leg. Grab PGA at knee convolutes and pull boots onto feet.	Point toes toward boots to avoid pickup of excess liner material.
	CAUTION	
	Do not sit on zippers.	
f.	Pull TLSA to knee position.	
g.	With front of TLSA hanging forward, go to squatting position placing TLSA front in lap.	
h.	Reach inside of right leg and connect UCTA quick disconnect to TLSA-UCTA hose assembly.	
i.	Aline index marks and connect biomedical harness electrical branch to suit electrical harness.	
j.	Verify that the communications carrier connection from electrical harness is pushed through neck ring.	
k.	Insert one arm completely into TLSA arm while simultaneously inserting head into neck ring with assistance from free arm. Insert free arm into other PGA arm.	Alternate methods, such as full insertion of arms into suit sleeves and then the insertion of head through neck ring, or insertion of one arm and then the combined insertion of second arm and head, are acceptable provided the load on the TLSA and/or the zippers is not increased.
l.	Stand erect to permit TLSA to settle over torso and limbs.	

4.1 PGA DONNING AND DOFFING

BASIC DATE May 1969   CHANGE DATE June 1971   PAGE 4-11

CSD-A-789-(2) III
APOLLO OPERATIONS HANDBOOK - EMU

STEP	PROCEDURE	REMARKS
m.	Close restraint and pressure-sealing zipper closures.	If unassisted, grasp back of neck ring and top of CWG with one hand. With other hand, grasp red donning lanyard, pull out and down to position restraint zipper slider toward middle of the back. Pull red donning lanyard over buttocks while running fingers ahead of slider to prevent snagging of CWG. Grasp cloth tab of restraint slider and close fully. Remove suit-donning lanyard, tuck slider tab up into TISA, and stow lanyard in pocket. Grasp back of neck ring with one hand and, with other hand, grasp blue donning lanyard. Pull lanyard out and down to fully close pressure-sealing zipper. Remove and stow blue suit-donning lanyard.
n.	Engage snap fastener of slider tab to receptacle on TISA.	
	Position lock assembly over slider and push lock button to LOCK.	
	Close cover flap over zipper and engage snap fasteners.	
o.	Don communications carrier by spreading the ear cups, and then make fine adjustments to the mike boom by bending it from the center.	Do not bend mike boom more than 80° from center. Bend mike boom from the middle section only. Do not straighten or bend the mike boom unless necessary.
	Adjust the chinstrap for proper fit.	

4.1 PGA DONNING AND DOFFING

BASIC DATE May 1969   CHANGE DATE June 1971   PAGE 4-12

APOLLO OPERATIONS HANDBOOK - EMU

STEP	PROCEDURE	REMARKS
p.	Connect electrical lead to upper branch of electrical harness.	
	CAUTION	
	Gas umbilicals must be inserted straight into the gas connectors to prevent excessive side loading and subsequent damage to gas connector O-ring seals.	
q.	Connect gas outlet umbilical to exhaust gas connector.	
	CAUTION	
	To prevent an inadvertent overpressure, always connect the outlet umbilical first.	
r.	Connect gas inlet umbilical to inlet gas connector.	
s.	Provide ventilation flow.	
4.1.1.5	Donning Pressure Helmet	Paragraphs 4.1.1.5 and 4.1.1.6 can be reversed provided that paragraph 4.1.1.5 is accomplished with decreased gas flow.
a.	Place helmet-attaching ring lock subassembly in ENGAGE.	
	CAUTION	
	PGA neck ring alinement marks must be alined with the helmet-engaged alinement marks to ensure proper helmet ventilation.	

4.1 PGA DONNING AND DOFFING

CSD-A-789-(2) III
APOLLO OPERATIONS HANDBOOK - EMU

STEP	PROCEDURE	REMARKS
b.	Place pressure helmet on helmet-attaching ring, ensure proper alinement by alining dual white lines, and press down on helmet.	
c.	Position lock subassembly to LOCK.	
4.1.1.6	Donning IV Gloves	
a.	Gloves or helmet may be difficult to don. If so, stop gas flow momentarily.	
b.	Place suit wrist disconnects in ENGAGE.	
c.	Ensure that glove palm restraint is loosened, then place hand into glove.	
d.	Aline glove wrist disconnect with suit wrist disconnect and engage both units.	
e.	Ensure that glove can be easily rotated and place the wrist disconnect to LOCK.	
f.	Adjust glove palm restraint.	
g.	Don other glove in the same way.	

4.1 PGA DONNING AND DOFFING

BASIC DATE May 1969   CHANGE DATE June 1971   PAGE 4-14

CSD-A-789-(2) III
APOLLO OPERATIONS HANDBOOK - EMU

STEP	PROCEDURE	REMARKS
4.1.2	PGA Doffing Procedures After IV Use	
4.1.2.1	Doffing IV Gloves	
a.	Decrease suit pressure to less than 0.75 psig.	
b.	Loosen palm restraint if required.	
c.	Place wrist disconnect in OPEN position and remove glove.	
d.	Doff other glove.	
4.1.2.2	Doffing Pressure Helmet	
a.	Depress tab, pull helmet-attaching ring latching mechanism, and rotate to the OPEN position.	
b.	Lift the helmet up and out of helmet-attaching ring.	
4.1.2.3	Doffing TLSA	See AOH LM and CSM procedures.
4.1.2.3.1	Doffing EV A7LB TLSA.-	See paragraph 4.1.2.3.2 for CMP A7LB doffing instructions.
a.	Empty UCTA.	
b.	Open slide fastener cover flap on ITMG.	
c.	Release the restraint zippers by pushing the lock on the restraint zipper lock assembly to the left and the red striker of the slider lock downward.	

4.1 PGA DONNING AND DOFFING

BASIC DATE May 1969   CHANGE DATE June 1971   PAGE 4-15

CSD-A-789-(2) III
APOLLO OPERATIONS HANDBOOK - EMU

STEP	PROCEDURE	REMARKS
d.	Grasp blue restraint zipper lanyard and pull upward until the zipper is fully open (pull lanyard under shoulder restraint cable as required). Grasp the blue waist zipper lanyard and pull around rear of suit to stop at left waist.	Disengage the donning aid hooks used to aline the vertical restraint zippers.
e.	With the right thumb, press downward on the arm assembly of the pressure-sealing zipper lock directly over the shaft and turn the arm clockwise with the index finger until the detent is engaged at the full-open position.	
f.	Grasp red pressure-sealing zipper lanyard and pull slider around rear waist to right side and fully upward to end of zipper at right shoulder (pull lanyard under shoulder restraint cable as required).	
g.	Disconnect gas inlet and outlet umbilicals.	
h.	Disconnect communications carrier electrical connector.	
i.	Grasp the TLSA at waist-entry area and separate the slide fasteners. Assume semierect position and slip the TLSA from around the buttocks and downward over legs.	
j.	If used, disconnect LCG multiple water connector, insert water connector plug into multiple water connector receptacle from the inside outward, and lock in place.	
k.	Grasp the helmet-attaching ring, slide head out, and pull both arms away from TLSA.	An alternate method of removing the TLSA by first removing one arm and then removing the head and other arm simultaneously is acceptable provided the load on the TLSA and/or zippers is not increased.

4.1 PGA DONNING AND DOFFING

BASIC DATE May 1969 CHANGE DATE June 1971 PAGE 4-16

CSD-A-789-(2) III
APOLLO OPERATIONS HANDBOOK - EMU

STEP	PROCEDURE	REMARKS
1.	Roll back cover to expose urine connector. Disconnect urine connector and biomedical connector.	
m.	Work the TLSA downward and remove legs from TLSA.	Don LCG booties after removing feet from TLSA. Remove LCG interconnect adapter and CWG electrical harness from stowage area. Close the restraint zipper before stowage.
4.1.2.3.2	Doffing CMP A7LB TLSA.-	
a.	Empty UCTA.	See AOH LM and CSM procedures.
b.	Open zipper-cover flap. Unlock pressure-sealing zipper lock and release slider tab snap fastener.	
c.	Fully open pressure and restraint zippers. CAUTION Do not sit on zippers.	If unassisted, remove blue donning lanyard from lanyard pocket and insert into pressure-sealing zipper tab. Grasp blue suit-donning lanyard, pull until the pressure-sealing zipper tab is fully OPEN, and remove suit-donning lanyard. Remove red suit-donning lanyard from lanyard pocket, release restraint zipper tab from stowed position, and insert lanyard in zipper tab. Grasp red suit-donning lanyard, pull until restraint zipper is full OPEN, and remove lanyard. Restow red and blue lanyards.

4.1 PGA DONNING AND DOFFING

BASIC DATE May 1969 CHANGE DATE June 1971 PAGE 4-17

CSD-A-789-(2) III
APOLLO OPERATIONS HANDBOOK - EMU

STEP	PROCEDURE	REMARKS
d.	Disconnect gas inlet and outlet hoses.	
e.	Disconnect communications carrier electrical lead and undo chinstrap to remove communications carrier. Remove communications carrier carefully. Do not bend mike boom.	
f.	Grasp PGA at rear entry area and separate slide fasteners; then slip TLSA from around back and buttocks.	
g.	Grasp helmet-attaching ring, slip head out, and pull both arms away from TLSA.	
h.	Disconnect biomedical connector.	
i.	Disconnect UCTA drain hose quick disconnect.	See paragraph 4.1.2.
j.	Remove legs from TLSA.	
k.	Replace protective covers and caps on TLSA electrical connector and two gas connectors.	
4.1.2.4	Doffing UCTA	
a.	Drain UCTA before doffing.	See AOH LM and CSM procedures.
b.	Partially doff PGA to disconnect the UCTA drain hose disconnect.	
c.	Disconnect UCTA drain hose from suit-mounted UCTA transfer hose assembly.	
d.	Complete doffing of PGA.	

4.1 PGA DONNING AND DOFFING

BASIC DATE May 1969   CHANGE DATE June 1971   PAGE 4-18

CSD-A-789-(2) III
APOLLO OPERATIONS HANDBOOK - EMU

STEP	PROCEDURE	REMARKS
e.	Remove UCTA elastic harness by detaching waistband Velcro patches while holding UCTA in place.	
f.	Remove cuff half of UCTA flange by using release button and twisting motion.	
g.	Roll cuff from penis.	
h.	Fold cuff and use UCTA clamp to close.	
4.1.2.5	Doffing Bioinstrumentation and CWG	To permanently remove bioinstrumentation system, remove bioinstrumentation system and cover exposed end of the PGA electrical umbilical and/or T-adapter cable with tape P/N SEB12100050-201 (on board).
a.	Disconnect electrical leads of biomedical electrode harness from signal conditioners in biomedical belt.	Use caution in mating or unmating biomedical connector to or from signal conditioners. Connectors are keyed for proper fit.
b.	Disengage snaps securing biomedical belt.	When transferring bioinstrumentation between CWG or LCG, or when reinstalling after removal, the signal conditioners should not be removed from the pockets in the biomedical belt. Only the nine-pin Airlock connector, the sternal electrode harness (blue code), and the auxiliary harness (yellow code) should be disconnected. The color

4.1 PGA DONNING AND DOFFING

CSD-A-789-(2) III
APOLLO OPERATIONS HANDBOOK - EMU

STEP	PROCEDURE	REMARKS
c.	Unbutton front opening.	code of the harness-to-signal conditioner should be observed when reinstalling the biomedical belt. The electrodes are <u>not</u> removed from skin to change garments.
d.	Pass bioinstrumentation electrode harness leads through a hole on chest area of CWG.	
e.	Remove CWG in the same manner as conventional long underwear.	
4.1.2.6	Doffing FCS	
a.	Doff FCS in the same manner as conventional underwear shorts.	
4.1.3	<u>PGA Donning Procedures Before EV Use</u>	
4.1.3.1	Donning FCS	
	See paragraph 4.1.1.1.	
	<u>NOTE</u>	
	Don UCTA first (para 4.1.1.3) when LCG is to be worn.	
4.1.3.2	Donning Bioinstrumentation and LCG or CWG	The LCG replaces the CWG for EVA. The LCG may be worn during periods of IVA. Verify that the UCTA hose is pulled completely through the LCG after donning LCG.

4.1 PGA DONNING AND DOFFING

BASIC DATE May 1969   CHANGE DATE June 1971   PAGE 4-20

CSD-A-789-(2) III
APOLLO OPERATIONS HANDBOOK - EMU

STEP	PROCEDURE	REMARKS
4.1.3.2.1	Donning Bioinstrumentation and LCG.-	The steps that prescribe the installation of the bioinstrumentation may be skipped when its use is not required (steps e through g, and j through m).
		The LCG will not interface with the CMP A7LB TLSA.
	CAUTION	
	Exercise care to avoid damaging LCG waterlines.	
a.	Unstow LCG from bag.	
b.	Open front entry fastener of LCG.	
c.	Don the LCG to waist as conventional long underwear.	
d.	Adjust feet into integrated socks.	
e.	Clean shaved electrode sites with alcohol and let dry.	See AOH LM procedures.
f.	Assemble signal conditioners, biomedical harness with biomedical belt, and test as a system.	Use caution in mating biomedical connector to signal conditioners. Connectors are keyed for proper fit.
g.	Insert bioinstrumentation leads through slots on chest area of LCG. Don top half of LCG, insert harness connector (blue and yellow) through holes in LCG. Snap biomedical belt to LCG.	When transferring bioinstrumentation between LCG or CWG or when reinstalling after removal, the signal conditioners should not be removed from the pockets in the biomedical belt. Only the nine-pin Airlock connector,

4.1 PGA DONNING AND DOFFING

CSD-A-789-(2) III
APOLLO OPERATIONS HANDBOOK - EMU

STEP	PROCEDURE	REMARKS
		the sternal electrode harness (blue code), and the auxiliary harness (yellow code) should be disconnected. The color code of the harness-to-signal conditioner should be observed when reinstalling the biomedical belt. The electrodes are not removed from skin to change garments.
h.	Complete LCG donning.	
i.	Close front entry.	
j.	Connect electrode harness to proper color-coded signal conditioner. Dot on electrode harness is facing out.	Verify that LCG manifold is outside of biomedical belt.
k.	Perform functional check.	
l.	Adjust ECG gain from normal signal to 40 percent of full scale.	
m.	Adjust ZPN gain to allow maximum inhalation to be at or near full scale.	
4.1.3.3	Donning TLSA	
4.1.3.3.1	Donning EV A7LB TLSA.- See paragraph 4.1.1.4.1.	
4.1.3.3.2	Donning CMP A7LB TLSA.- See paragraph 4.1.1.4.2.	

4.1 PGA DONNING AND DOFFING

BASIC DATE May 1969   CHANGE DATE June 1971   PAGE 4-22

CSD-A-789-(2) III
APOLLO OPERATIONS HANDBOOK - EMU

STEP	PROCEDURE	REMARKS
4.1.3.4	Donning Lunar Boots	
a.	Insert PGA boots into lunar boots and position with attached donning straps.	
b.	Engage snap fasteners on tongue of boot.	
c.	Latch adjustment strap and buckle.	
4.1.3.5	Donning PLSS/OPS	
	See paragraph 4.3.1.	
4.1.3.6	Donning Pressure Helmet	See paragraph 4.10.1.2.a.
a.	Apply antifog solution.	
b.	Helmet may be difficult to don. If so, stop gas flow momentarily.	
	CAUTION	
	Prior to donning helmet, PGA diverter valves must be EV (vertical) position if O$_2$ flow is to be provided by PLSS/OPS.	
c.	Place helmet-attaching ring lock subassembly in ENGAGE.	
d.	Place pressure helmet on helmet-attaching ring, ensure proper alinement by alining dual white lines, and press down on helmet.	

4.1 PGA DONNING AND DOFFING

STEP	PROCEDURE	REMARKS
e.	Position lock subassembly to LOCK.	
	CAUTION	
	PGA neck ring alinement marks must be alined with the helmet-locked alinement marks to ensure proper helmet ventilation.	
4.1.3.7	Donning EV Gloves	
a.	Don comfort gloves and wristlets if desired.	
b.	Roll glove gauntlet back to provide access to wrist disconnect.	
c.	Place suit wrist disconnect in ENGAGE.	
d.	Gloves may be difficult to don. If so, stop gas flow momentarily.	
e.	Loosen palm restraint if necessary and place hand into glove.	
f.	Aline glove wrist disconnect with suit wrist disconnect and engage both units.	
g.	Ensure gloves rotate easily and place suit wrist disconnects in their LOCK position.	
h.	Roll glove gauntlet back over wrist ring.	
i.	Adjust palm restraint as desired. Close cover flap and engage fasteners.	

4.1 PGA DONNING AND DOFFING

CSD-A-789-(2) III
APOLLO OPERATIONS HANDBOOK - EMU

STEP	PROCEDURE	REMARKS
4.1.3.8	Donning LEVA	LEVA may be installed on helmet before donning helmet, thereby allowing LEVA and helmet to be donned as a unit.
a.	Verify that visors are open (up).	
b.	Disengage latching mechanism through access on LEVA collar.	Raise LEVA collar prior to helmet donning, then don helmet with LEVA and lower collar to cover neckring and fasten front and back.
c.	Place LEVA over pressure helmet and lower onto helmet-attaching ring.	
d.	Aline separation of plastic collar with helmet-engaged alinement marks.	Aline LEVA by using the projecting PGA feedport as a guide.
e.	Ensure LEVA is properly located on attaching ring and lock.	
f.	Lower collar to cover neckring and fasten front and back.	
	CAUTION	
	Collar must conceal helmet-attaching ring area for lunar surface activity only.	
4.1.3.9	Donning BSLSS	
	See paragraph 4.11.3.	
4.1.4	PGA Doffing Procedures After EV Use	The EMU is assumed to be in the same configuration as at end of extravehicular configuration donning procedures.

4.1 PGA DONNING AND DOFFING

BASIC DATE _____ May 1969 _____ CHANGE DATE June 1971 PAGE 4-25

APOLLO OPERATIONS HANDBOOK - EMU

4.1 PGA DONNING AND DOFFING

STEP	PROCEDURE	REMARKS
4.1.4.1	Disconnecting BSLSS	
	See paragraph 4.11.7	
4.1.4.2	Doffing EV Gloves	
a.	Decrease PGA pressure flow to 0.75 psig or less.	
b.	Roll glove gauntlet back and put wrist disconnect in OPEN position.	
c.	Doff glove.	
d.	Similarly doff other glove.	
4.1.4.3	Doffing LEVA	LEVA may be doffed with helmet as a unit. Raise LEVA collar and remove helmet per paragraph 4.1.4.4.
a.	Verify both visors in full OPEN (up) position.	
b.	Disengage fastener tapes of LEVA collar.	
c.	Disengage locking mechanism.	
d.	Ease LEVA up and off pressure helmet.	
4.1.4.4	Doffing Helmet	
a.	Depress tab, pull helmet-attaching ring latching mechanism and rotate to the OPEN position.	
b.	Ease pressure helmet up and out of helmet-attaching ring.	

BASIC DATE May 1969 CHANGE DATE June 1971 PAGE 4-26

CSD-A-789-(2) III
APOLLO OPERATIONS HANDBOOK - EMU

STEP	PROCEDURE	REMARKS
4.1.4.5	Doffing PLSS/OPS	
	See paragraphs 4.3.2 and 4.3.3	
4.1.4.6	Doffing Lunar Boots	
	a. Unbuckle adjusting strap and unsnap fasteners.	
	b. Slip boots off.	
4.1.4.7	Doffing TLSA	
	See paragraph 4.1.2.3.	
4.1.4.7.1	Doffing EV A7LB TLSA.- (See paragraph 4.1.2.3.1.)	
4.1.4.7.2	Doffing CMP A7LB TLSA.- (See paragraph 4.1.2.3.2.)	
4.1.4.8	Doffing UCTA	
	See paragraph 4.1.2.4.	
4.1.4.9	Doffing Bioinstrumentation and CWG or LCG	To permanently remove bioinstrumentation system, remove bioinstrumentation system and cover exposed end of the PGA electrical umbilical and/or T-adapter cable with tape P/N SEB12100050-201 (on board).
4.1.4.9.1	Doffing LCG.-	
	a. Disconnect electrical leads of biomedical electrode harness from signal conditioners in biomedical belt.	Use caution in mating or unmating biomedical connector to or from signal conditioners. Connectors are keyed for proper fit.
	b. Disengage snaps securing biomedical belt.	

4.1 PGA DONNING AND DOFFING

CSD-A-789-(2) III
APOLLO OPERATIONS HANDBOOK-EMU

STEP	PROCEDURE	REMARKS
		When transferring bioinstrumentation between CWG or LCG, or when reinstalling after removal, the signal conditioners should not be removed from the pockets in the biomedical belt. Only the nine-pin Airlock connector, the sternal electrode harness (blue code), and the auxiliary harness (yellow code) should be disconnected. The color code of the harness-to-signal conditioner should be observed when reinstalling the biomedical belt. The electrodes are not removed from skin to change garments.
c.	Open front entry.	
d.	Pass bioinstrumentation electrode harness leads through holes in chest area of LCG.	
e.	Remove LCG in the same manner as conventional long underwear.	
4.1.4.9.2	Doffing CWG.- (See paragraph 4.1.2.5.)	
4.1.4.10	Doffing FCS	
	See paragraph 4.1.2.6.	

4.1 PGA DONNING AND DOFFING

APOLLO OPERATIONS HANDBOOK - EMU

STEP	PROCEDURE	REMARKS
4.2	OPS OPERATIONS	
4.2.1	OPS Checkout	
a.	Open access flaps and verify OPS bottle pressure gage reads 5880 ± 500 psia.	
b.	Verify OPS O_2 connector locked in stowage plate.	
c.	Set OPS actuation lever to ON.	
d.	Verify OPS regulator checkout gage reads 3.70 ± 0.30 psid.	
e.	Set OPS O_2 actuation lever to OFF.	The OPS regulator checkout gage will continue to read 3.7 ± 0.3 psid for approximately 3 minutes after OPS actuation lever is OFF.
f.	Secure all access flaps.	
g.	Verify OPS regulator checkout gage less than 2.5 psi.	
4.2.2	OPS Donning Prior to Contingency Transfer	
a.	Pull out tear-tack stitches on the PLSS adjustable harness by yanking on end tab until strap is free.	
b.	Open thermal cover over strap buckle adjustment and remove retaining spring clip.	

4.2 OPS OPERATIONS

BASIC DATE May 1969 CHANGE DATE June 1971 PAGE 4-29

CSD-A-789-(2) III
APOLLO OPERATIONS HANDBOOK - EMU

STEP	PROCEDURE	REMARKS
c.	Join the fixed waist harness to the adjustable waist harness of the PLSS using the D-buckle under the thermal cover of the adjustable harness (PLSS attachment end) and the hook under the thermal cover on the fixed length strap (PLSS attachment end). The hook on the free end of the straps should be faced inward.	
d.	Loop harnesses around the back of PGA and thread through LM tether restraints.	
e.	Remove OPS O_2 connector from stowage plate.	
f.	Hook harnesses to OPS.	
g.	Install OPS oxygen connector (blue) into one of the PGA inlet oxygen connectors (blue) and verify locked.	
h.	Install purge valve into one of the PGA outlet oxygen connectors (red) and verify locked.	
i.	Open OPS actuator access flap.	
j.	Adjust harnesses to secure OPS. Allow for the expansion of the PGA when pressurized.	
4.2.3	OPS Donning Before CMP EVA	
a.	Perform OPS checkout per paragraph 4.2.1.	
b.	Verify OPS pressure indicated on checkout gage is less than 2.5 psid.	

4.2 OPS OPERATIONS

BASIC DATE May 1969   CHANGE DATE June 1971   PAGE 4-30

CSD-A-789-(2) III
APOLLO OPERATIONS HANDBOOK - EMU

STEP	PROCEDURE	REMARKS
c.	Unstow OPS O$_2$ connector.	
d.	Attach straps (4) to the OPS.	
e.	Attach OPS adapter plate to PGA upper PLSS bracket.	
f.	Don OPS by attaching the OPS bottom straps (2) to the PGA D-ring and the OPS top straps (2) to the adapter plate.	Left OPS strap attaches to the left side of the adapter plate and right OPS strap attaches to the right side of the adapter plate.
g.	Route OPS actuator cable over right shoulder and back to adapter plate.	
h.	Route OPS O$_2$ hose under OPS and under left arm. Connect to the left PGA inlet O$_2$ connector (blue).	
i.	Verify hose connector is locked.	

4.2 OPS OPERATIONS

BASIC DATE May 1969   CHANGE DATE June 1971   PAGE 4-31

CSD-A-789-(2) III
APOLLO OPERATIONS HANDBOOK - EMU

STEP	PROCEDURE	REMARKS
4.2.4	OPS Activation	
	Move OPS actuation lever from OFF to ON position (pull down, rotate up) and allow it to lock.	
4.2.5	OPS Deactivation	
	Move OPS actuation lever from ON to OFF position (pull up, rotate down) and allow it to lock.	
4.2.6	OPS Doffing After CMP EVA	
a.	Disconnect OPS actuator from adapter plate.	
b.	Disconnect OPS O_2 hose connector from PGA.	
c.	Disconnect straps (4) from adapter plate and PGA D-ring.	
d.	Remove OPS adapter plate and stow.	

4.2 OPS OPERATIONS

BASIC DATE May 1969   CHANGE DATE June 1971   PAGE 4-32

CSD-A-789-(2) III
APOLLO OPERATIONS HANDBOOK - EMU

STEP	PROCEDURE	REMARKS
4.3	PLSS/OPS DONNING AND DOFFING	
4.3.1	PLSS/OPS Donning	Helmet and gloves off. PGA donned. PGA diverter valves (2) horizontal.
a.	Open access flap, unstow antenna connector, OPS half, and secure access flaps.	
b.	Lift OPS locking pin.	
c.	Slide OPS onto PLSS from left to right while facing PLSS conformal side.	Conformal side of PLSS is the side that conforms to the crewman's back when PLSS is donned.
d.	Push locking pin down.	
e.	Remove EVCS antenna connector (J5) dust cap.	
f.	Connect antenna connector OPS half to antenna connector PLSS half. Screw on CW.	
g.	Verify sublimator exhaust is clear.	Visual inspection.
h.	Unstow PLSS shoulder and waist harnesses.	
i.	Unstow PLSS PGA electrical umbilical, inlet and outlet O_2, and multiple water connectors.	
j.	Remove battery cable from stowage plate. Rotate battery cable handle CCW (90° to alinement marks).	
k.	Remove battery connector (J6) dust cap and stow on battery cable stowage plate.	

4.3 PLSS/OPS DONNING AND DOFFING

BASIC DATE May 1969   CHANGE DATE June 1971   PAGE 4-33

CSD-A-789-(2) III
APOLLO OPERATIONS HANDBOOK - EMU

STEP	PROCEDURE	REMARKS
1.	Connect battery cable to battery.	
1.	Aline marks on battery cable body and handle.	
2.	Aline marks on battery cable handle and battery.	
3.	Connect battery cable to battery connector and rotate handle CW (90°).	
m.	Remove dust cap RCU connector (J3), PLSS half. Twist CCW.	
n.	Verify OPS actuation lever is OFF and OPS regulator check-out gage reads less than 2.5 psi.	Do not unstow OPS oxygen hose at this time.
o.	Unstow OPS connector. Pull stowage plate tabs and rotate CCW.	
p.	Secure PLSS/OPS access flaps.	
q.	Don PLSS/OPS by securing shoulder and waist harnesses to the PGA upper and lower PLSS brackets.	
r.	Connect PLSS O_2 outlet (blue) to inboard PGA inlet connector (blue) and PLSS O_2 inlet (red) to upper PGA outlet connector (red). Connect PLSS multiple water connector to PGA multiple water connector (blue). Connect PLSS/PGA electrical umbilical and lock.	
s.	Before connecting RCU, verify: PLSS pump switch — OFF, PLSS fan switch — OFF, and PLSS mode selector switch — position 0 (OFF).	

4.3 PLSS/OPS DONNING AND DOFFING

BASIC DATE May 1969   CHANGE DATE June 1971   PAGE 4-34

CSD-A-789-(2) III
APOLLO OPERATIONS HANDBOOK - EMU

STEP	PROCEDURE	REMARKS
t.	Connect RCU electrical connector to the PLSS.	
1.	Aline marks on RCU connector body and handle.	
2.	Aline marks on RCU connector and PLSS, insert, and rotate CW 90°.	
u.	Attach RCU to PLSS straps and PGA as follows:	
1.	Pull Velcro strap away from front of RCU.	
2.	Using strap as a grip, pull directly forward of RCU and then down. Release strap to lock in open position.	
3.	Insert lower clip to PGA upper PLSS bracket.	
4.	Raise RCU and insert the left shoulder clip into left RCU clip, then clip the right side.	
5.	To lock clips, pull strap handle forward and up to a horizontal position and release. Verify hooks are locked in closed position.	
6.	Restow Velcro strap handle on front of RCU.	
v.	Unstow OPS hose.	
w.	Depress OPS actuation lever bracket tab and unstow actuation lever cable.	

4.3 PLSS/OPS DONNING AND DOFFING

BASIC DATE May 1969   CHANGE DATE June 1971   PAGE 4-35

CSD-A-789-(2) III
APOLLO OPERATIONS HANDBOOK - EMU

STEP	PROCEDURE	REMARKS
x.	Attach OPS actuation lever to the RCU.	
1.	Insert lower pins on the OPS actuation lever into the RCU actuation lever bracket slots.	
2.	Push upper portion of the OPS actuation lever toward the bracket until the upper pins engage the bracket and snap is locked.	
y.	Install OPS connector to the unused PGA connector (blue to blue) and lock.	
z.	Install purge valve in unused PGA O_2 connector (red to red).	
aa.	Secure all PLSS/OPS access flaps and verify gas connector lock locks (4).	
ab.	Unstow antenna.	
4.3.2	PLSS/OPS Doffing in Pressurized Environment	a. Helmet and gloves off.
		b. PLSS primary and auxiliary feedwater valves closed (up).
		c. PLSS O_2 shutoff valve off (up).
		d. LM is at 5 psia.
		e. OPS actuation lever off.

4.3 PLSS/OPS DONNING AND DOFFING

BASIC DATE May 1969 CHANGE DATE June 1971 PAGE 4-36

CSD-A-789-(2) III
APOLLO OPERATIONS HANDBOOK - EMU

STEP	PROCEDURE	REMARKS
a.	Remove OPS actuation lever from RCU.	
b.	Disconnect RCU from PGA upper PLSS bracket and PLSS shoulder harnesses.	
1.	Pull Velcro strap away from front of RCU.	
2.	Using strap as a grip, pull directly forward of RCU and then down. Release strap to lock in open position.	
3.	Lift RCU from left shoulder strap, then right.	
4.	Lower RCU from PGA upper PLSS bracket.	
	CAUTION Before disconnecting RCU, all electrical PLSS controls must be OFF. PLSS pump switch — OFF PLSS fan switch — OFF PLSS mode selector switch — Position 0 (OFF)	
c.	Disconnect RCU electrical umbilical from PLSS by rotating RCU connector handle CCW (90°).	
d.	Disconnect inlet and outlet O_2 and multiple water connectors, electrical umbilical connector, and OPS O_2 connector. The electrical umbilical connector is pulled away from the PGA and rotated CCW to remove.	

4.3 PLSS/OPS DONNING AND DOFFING

BASIC DATE May 1969   CHANGE DATE June 1971   PAGE 4-37

CSD-A-789-(2) III
APOLLO OPERATIONS HANDBOOK - EMU

STEP	PROCEDURE	REMARKS
e.	Remove the purge valve and stow.	
f.	Remove PLSS shoulder and waist harnesses from the PGA and doff the PLSS.	
g.	Temporarily stow the PLSS.	
h.	Stow OPS antenna.	
i.	Stow OPS actuation lever.	
1.	Insert lower pins on the OPS actuation lever into the OPS actuation lever bracket slots.	
2.	Push upper portion of the OPS actuation lever down until the upper pins engage the bracket and snap lock in place.	
j.	Verify OPS actuation lever is OFF and locked.	
k.	Stow OPS O$_2$ hose and connector; secure OPS access covers.	
1.	Route the hose around the back of the OPS over the actuation lever cable.	
2.	Verify the OPS connector stowage plate is in the open position (CCW).	
3.	Insert the OPS connector and lock the OPS stowage plate connector (twist CW).	
4.	Secure access flaps over OPS hose.	

4.3 PLSS/OPS DONNING AND DOFFING

BASIC DATE May 1969 CHANGE DATE June 1971 PAGE 4-38

CSD-A-789-(2) III
APOLLO OPERATIONS HANDBOOK - EMU

STEP	PROCEDURE	REMARKS
1.	Stow PLSS inlet and outlet O_2 and multiple water hoses and PLSS PGA electrical umbilical and connector.	
l.	Confirm O_2, water, and electrical connector stowage plates are in the open position (CCW).	
2.	Perform stowage routing according to the decal on the PLSS O_2 bottle shield.	Refer to hose routing decal on conformal side (against crewman's back) of PLSS.
3.	Lock all stowage connectors (twist CW).	
m.	Disconnect OPS antenna connector from EVCS by unscrewing CCW.	
n.	Replace antenna connector dust cap by pushing straight on.	
o.	Lift OPS locking pin to release.	
p.	Slide OPS off PLSS from right to left while facing PLSS conformal side.	Conformal side of the PLSS is the side that conforms to the crewman's back when PLSS is donned.
q.	Stow antenna connector inside OPS by screwing antenna connector CW on the stowage plate.	
r.	Secure OPS access covers.	
s.	Replace RCU connector (J3) dust cap by alining marks, inserting, and twisting CW.	
t.	Restow PLSS shoulder and waist harnesses.	

4.3 PLSS/OPS DONNING AND DOFFING

BASIC DATE May 1969   CHANGE DATE June 1971   PAGE 4-39

CSD-A-789-(2) III
APOLLO OPERATIONS HANDBOOK - EMU

STEP	PROCEDURE	REMARKS
4.3.3	PLSS/OPS Doffing in Unpressurized Environment	
a.	Verify OPS actuation lever OFF.	
b.	Disconnect purge valve and stow.	
c.	Disconnect OPS O_2 connector.	
d.	Connect ECS O_2 umbilicals to PGA (red to red, blue to blue); lock and actuate ECS.	
e.	Set PLSS fan switch OFF.	
f.	Set PLSS O_2 shutoff valve OFF (up).	O_2 shutoff handle safety must be depressed as handle is pulled forward.
g.	Verify PGA pressure gage reads 3.6 to 4.3 psi. Turn PGA diverter valves (2) horizontal.	
h.	Set PLSS mode selector switch to position 0 (OFF).	
i.	Disconnect PLSS PGA electrical umbilical. The electrical umbilical connector is pulled away from the PGA and rotated CCW to remove. Connect LM communication umbilical.	
j.	Verify PLSS primary and auxiliary feedwater valves CLOSED (up).	
k.	Verify PLSS pump switch OFF.	
l.	Remove OPS actuation lever from the RCU. Depress release lever to free actuation lever upper pins and remove.	

4.3 PLSS/OPS DONNING AND DOFFING

BASIC DATE May 1969   CHANGE DATE June 1971   PAGE 4-40

CSD-A-789-(2) III
APOLLO OPERATIONS HANDBOOK - EMU

STEP	PROCEDURE	REMARKS
m.	Disconnect RCU from PGA upper PLSS bracket and PLSS shoulder harnesses.	
1.	Pull Velcro strap away from front of RCU.	
2.	Using strap as a grip, pull directly forward of RCU and then down. Release strap to lock in open position.	

4.3 PLSS/OPS DONNING AND DOFFING

CSD-A-789-(2) III
APOLLO OPERATIONS HANDBOOK - EMU

STEP	PROCEDURE	REMARKS
3.	Lift RCU from left shoulder strap, then right.	
4.	Lower RCU from PGA upper PLSS bracket.	
	CAUTION	
	Before electrically disconnecting RCU, all PLSS electrical controls must be in OFF position.	
	PLSS pump switch OFF	
	PLSS fan switch OFF	
	PLSS mode selector switch Position 0 (OFF)	
n.	Disconnect RCU electrical connector from the PLSS. Rotate RCU electrical connector handle CCW to disconnect.	
o.	Disconnect PLSS inlet and outlet O_2 and multiple water connectors. Disconnect outlet O_2 connector (red) first.	
p.	Remove waist harnesses from the PGA.	Crewman will require assistance.
1.	Grasp outside loop of right-hand strap between adjustment buckle and PGA hook, and tear tack-stitches adjacent to buckle.	
2.	Unsnap harness keeper between adjustment buckle and PLSS hook.	
3.	Grasp exposed end of strap between PLSS hook and adjustment buckle, and tear tack-stitches.	

4.3 PLSS/OPS DONNING AND DOFFING

BASIC DATE May 1969   **CHANGE DATE** June 1971   **PAGE** 4-42

CSD-A-789-(2) III
APOLLO OPERATIONS HANDBOOK - EMU

STEP	PROCEDURE	REMARKS
4.	Unsnap adjustment-buckle thermal insulation to expose buckle.	
5.	Grasp buckle roller release tab and rotate outward to release grip on harness. Lengthen the harness by use of the adjustment buckle.	
6.	Unhook right-hand harness from PGA.	
7.	Unhook left-hand harness from PGA.	
q.	Remove shoulder harnesses from PGA and doff PLSS/OPS.	
r.	Temporarily stow PLSS/OPS.	
s.	Stow antenna.	
t.	Replace RCU connector (J3) dust cap by alining marks, inserting, and twisting CW.	
u.	Stow PLSS inlet and outlet O_2 hoses and connectors, multiple water hoses and connectors, and PLSS-PGA electrical umbilical in the stowage connectors provided. Refer to hose routing decal on conformal side of PLSS. Verify all connectors locked in place. Secure hoses with hose stowage strap.	
v.	Stow PLSS straps.	
w.	Disconnect OPS antenna connector from EVCS. Unscrew CCW.	
x.	Replace antenna connector dust cap by pushing into place.	
y.	Secure all PLSS thermal flaps.	

4.3 PLSS/OPS DONNING AND DOFFING

BASIC DATE May 1969   CHANGE DATE June 1971   PAGE 4-43

STEP	PROCEDURE	REMARKS
z.	Stow OPS actuation lever.	
1.	Insert lower pins on OPS actuation lever into OPS actuation lever bracket slots.	
2.	Push upper portion of OPS/actuation lever down until the upper pins engage the bracket and lock in place.	
3.	Verify actuation lever is off and locked.	
aa.	Stow OPS hose and connector, and secure OPS thermal flaps.	Hose is routed around the back of the OPS, over the actuator cable, and is held in place by the thermal cover. The connector stowage plate must be rotated full CCW to open before the connector is inserted, and then rotated CW to lock the connector in place.

4.3 PLSS/OPS DONNING AND DOFFING

CSD-A-789-(2) III
APOLLO OPERATIONS HANDBOOK - EMU

STEP	PROCEDURE	REMARKS
4.4	PLSS COMMUNICATIONS CHECK	Crewmen suited with helmets off. Vent flow provided by LM ECS. OPS, RCU, PLSS, and PGA systems are properly connected. PLSS/EVCS modes interface with LM and CM communications subsystems. Spacecraft switch positions for various communications modes (PLSS, SC, MSFN) are found in the AOH for CSM and LM. PLSS switches and valves off.
a.	Set PLSS mode selector switch to position B.	
b.	Verify 1.5 kHz warble tone on for 10 seconds. Low-vent flow warning flag shows P, low PGA pressure warning flag shows 0.	
c.	Read PLSS O_2 gage (percent of full scale).	
d.	Verify voice communications. Adjust volume.	Increase volume by rotating blade CCW.
e.	Set PLSS mode selector switch to position A.	
f.	Verify 1.5 kHz warble tone on for 10 seconds. Low-vent flow warning flag continues to show P. Low PGA pressure warning flag continues to show 0.	
g.	Read PLSS O_2 gage (percent of full scale).	
h.	Verify voice communications and TM, adjust volume.	Increase volume by rotating wheel CCW.
i.	PLSS mode selector switch to position AR.	
j.	Verify 1.5 kHz warble tone on for 10 seconds. Low-vent flow warning flag shows P. Low PGA pressure warning flag shows 0.	

4.4 PLSS COMMUNICATIONS CHECK

BASIC DATE May 1969 **CHANGE DATE** June 1971 **PAGE** 4-45

CSD-A-789-(2) III
APOLLO OPERATIONS HANDBOOK - EMU

STEP	PROCEDURE	REMARKS
k.	Read PLSS O_2 gage (percent of full scale).	
l.	Verify voice communication and TM; adjust volume as required.	

4.4 PLSS COMMUNICATIONS CHECK

BASIC DATE May 1969   CHANGE DATE June 1971   PAGE 4-46

APOLLO OPERATIONS HANDBOOK - EMU

STEP	PROCEDURE	REMARKS
4.5	PRESSURE INTEGRITY CHECK	
a.	Set PLSS O_2 shutoff valve to ON (down).	EMU donned. Ambient pressure 5.0 psia. PLSS fan in ON. PLSS primary and auxiliary valves CLOSED.
b.	Verify PLSS 1.5 kHz warble tone on for 10 seconds. High O_2 flow warning flag shows 0. Low PGA pressure warning flag shows 0 and clears when PGA pressure reaches 3.1 to 3.4 psid.	
c.	Verify high O_2 flow warning flag clears as PGA gage reaches 3.85 ± 0.15 psig.	
d.	Set PLSS O_2 shutoff valve to OFF (up).	O_2 shutoff handle safety must be depressed as handle is pulled forward.
e.	Read PGA pressure gage and monitor pressure decay for 1 minute.	
f.	Report pressure decay.	This step is not considered as a go/no-go check, but is used primarily as a gross leak check. If pressure decay exceeds 0.3 psid, all gas connectors, neck ring, and glove connectors should be verified locked.
g.	Set PLSS O_2 shutoff valve to ON (down).	Warning tone and Hi O_2 Flow flag may come on.

4.5 PRESSURE INTEGRITY CHECK

BASIC DATE May 1969   CHANGE DATE June 1971

APOLLO OPERATIONS HANDBOOK - EMU

STEP	PROCEDURE	REMARKS
h.	Verify PGA pressure is 3.85 ± 0.15 psi and all warning flags are clear.	Clearing of the "Hi Flow" warning flag indicates leakage plus usage is less than 0.28 lb/hr when PGA pressure is at 4 psia.

4.5 PRESSURE INTEGRITY CHECK

APOLLO OPERATIONS HANDBOOK - EMU

STEP	PROCEDURE	REMARKS
4.6	PLSS NORMAL OPERATING MODES	
4.6.1	PLSS Activation	
a.	Set PLSS mode selector switch to position AR.	PLSS/OPS donned; helmet and gloves off.
b.	Set PLSS fan switch to ON.	A 1.5-kHz warble tone on for 10 seconds. Low-vent flow warning flag shows P. Low PGA pressure warning flag shows O. Verify communication and TM.
c.	Don helmet and gloves.	Verify low-vent flow warning flag clear. If fan is activated for more than 30 minutes without PGA cooling, visor fogging may occur.
d.	Set PLSS O_2 shutoff valve to ON (down).	
	1. Verify low PGA pressure warning flag clear.	
	2. Verify 1.5 kHz tone for 10 seconds and warning flag shows O, and then clear when PGA reaches high O_2 flow 3.85 ± 0.15 psig.	
e.	Set PLSS pump switch to ON.	Low feedwater pressure warning tone on and warning flag indicates A between 1.2 and 1.7 psia cabin pressure.
f.	Verify diverter valve in the MINIMUM position.	
g.	Set PLSS feedwater valve to OPEN (down).	Ambient pressure must be below 1000μ Hg before opening valve.

4.6 PLSS NORMAL OPERATING MODES

CSD-A-789-(2) III
APOLLO OPERATIONS HANDBOOK - EMU

STEP	PROCEDURE	REMARKS
h.	Position PLSS H_2O diverter valve for comfort after low feedwater pressure warning flag clears.	
4.6.2	Wet Sublimator Restart	
a.	Verify PLSS primary and auxiliary feedwater valve is CLOSED (up).	
b.	Verify PLSS H_2O diverter valve is at MAXIMUM (down).	PLSS operating. Ambient pressure at vacuum.
c.	Maintain workload to deplete feedwater rapidly.	
d.	Verify 1.5-kHz warble tone for 10 seconds. Low feedwater pressure warning flag shows A.	
e.	After 5 minutes, set PLSS H_2O diverter valve to MINIMUM (up).	
f.	Set PLSS primary feedwater valve to OPEN (down). Select desired diverter position after low feedwater pressure warning flag clears.	
4.6.3	PLSS Deactivation	
a.	Set PLSS primary and auxiliary water feedwater valve to CLOSED (up).	EMU donned, PLSS operating. Ambient pressure at vacuum.
b.	Repressurize LM cabin.	This step is performed prior to repressurization to prevent loss of feedwater when pressure is reestablished.
c.	Set PLSS O_2 shutoff valve to OFF (up).	
d.	Set pump switch to OFF. PGA pressure is equalized with ambient. Helmet and gloves are doffed.	

4.6 PLSS NORMAL OPERATING MODES

BASIC DATE May 1969   CHANGE DATE June 1971   PAGE 4-50

CSD-A-789-(2) III
APOLLO OPERATIONS HANDBOOK - EMU

STEP	PROCEDURE	REMARKS
e.	Set fan switch to OFF.	
f.	Set PLSS mode-selector switch to position 0 (OFF).	

4.6 PLSS NORMAL OPERATING MODES

BASIC DATE May 1969   CHANGE DATE June 1971   PAGE 4-51

STEP	PROCEDURE	REMARKS
4.7	PLSS RECHARGE AND FEEDWATER REMOVAL PROCEDURES	The PLSS recharge procedures consist of battery replacement, LiOH cartridge change, oxygen system recharge, and feedwater reservoir recharge. Oxygen and feedwater recharge procedures are given in the LM AOH.
4.7.1	Battery Replacement	
a.	If RCU is connected electrically to the PLSS, all electrical controls must be in OFF position before connecting or disconnecting battery cable. PLSS pump switch OFF PLSS fan switch OFF PLSS mode selector switch position 0 (OFF)	
b.	Rotate PLSS main battery cable 90° CCW and remove from battery. Remove protective cover from main battery cable stowage connector and stow on battery (pull knob on battery locking device outward and slide down to unlock battery).	
c.	Remove old battery from PLSS and stow.	
d.	Obtain replacement battery, remove dust cap, aline battery on battery foot, and slide into place in PLSS.	
e.	Pull knob on battery locking device outward and slide up to lock battery.	
f.	Connect PLSS main battery cable to battery.	

4.7 PLSS RECHARGE AND FEEDWATER REMOVAL PROCEDURES

CSD-A-789-(2) III
APOLLO OPERATIONS HANDBOOK - EMU

STEP	PROCEDURE	REMARKS
g.	Verify battery lift strap snapped in a loop.	
h.	Insert PLSS hose stowage strap through lift strap loop and stow.	
4.7.2	Removal of LiOH Cartridge From Stowage Container	
a.	Verify that green marking on indicator pin is visible on cover of stowage container.	Decal: 110°F 120°F 130°F 140°F Center turns black at rating shown
b.	Lift ring and remove tape from relief valve.	
c.	Depress relief valve button until indicator pin retracts.	Decal: **CAUTION** Do not use cartridge if green marking on indicator pin is not visible. TO OPEN: 1. Lift ring and remove tape from relief valve. 2. Depress relief valve button until indicator pin retracts. 3. Pull ring to remove locking hooks. 4. Lift off cover.
d.	Pull ring to remove locking hooks.	
e.	Lift off cover.	
f.	Check temperature indicator on end of LiOH cartridge.	
g.	Verify that 130° F temperature monitor dot is white.	

4.7 PLSS RECHARGE AND FEEDWATER REMOVAL PROCEDURES

BASIC DATE May 1969   CHANGE DATE June 1971   PAGE 4-53

CSD-A-789-(2) III
APOLLO OPERATIONS HANDBOOK - EMU

STEP	PROCEDURE	REMARKS
4.7.3	<u>LiOH Cartridge Change</u>	
a.	Verify PLSS O_2 shutoff valve is OFF.	
b.	Remove thermal insulation from the canister cover.	
c.	Depress cover lock.	
d.	Rotate canister cover CCW until alinement mark on cover is alined with the open mark on canister.	
e.	Remove cover by pulling from canister.	
f.	Grasp drop handle and rotate contaminant control cartridge CCW until alinement marks on cartridge and canister assembly are alined.	
g.	Pull spent contaminant cartridge out of canister.	
h.	Obtain replacement cartridge, grasp drop handle, aline marks, and insert replacement cartridge into canister until it bottoms.	
i.	Rotate cartridge CW approximately 90° until marks are alined to lock into position.	
j.	Ascertain that alinement marks on both parts of the cover are alined. Grasp cover by handle and depress cover lock.	
k.	Aline the alinement marks on canister cover with the "open" mark on cover.	

4.7 PLSS RECHARGE AND FEEDWATER REMOVAL PROCEDURES

BASIC DATE May 1969   CHANGE DATE June 1971   PAGE 4-54

4.7 PLSS RECHARGE AND FEEDWATER REMOVAL PROCEDURES

STEP	PROCEDURE	REMARKS
l.	Insert cover in canister.	
	CAUTION	
	Do not force cover into canister if slightly misalined as this may damage cover seal. First, aline cover properly.	
m.	Rotate cover CW until alinement mark on cover is alined with "closed" mark on canister.	
n.	Resnap insulation flap over canister cover.	

CSD-A-789-(2) III
APOLLO OPERATIONS HANDBOOK - EMU

STEP	PROCEDURE	REMARKS
4.8	PURGE VALVE OPERATION	
4.8.1	Activation Procedures for Purge Valve	
a.	Remove pull pin by grasping the red apple and pulling with about 20 pounds of force.	
b.	Squeeze the two locktabs on the purge valve barrel simultaneously. The valve will now pop open.	
4.8.2	Flow Adjustment Procedures	
a.	To adjust from HIGH flow to LOW flow, depress gold button on face of purge valve and rotate until it stops (45°) in direction indicated on purge valve face (top of flange moves to crewman's left).	
b.	To adjust from LOW flow to HIGH flow, depress gold button on face of purge valve and rotate until it stops (45°) in direction indicated on purge valve face (top of flange moves to crewman's right).	
4.8.3	Shutoff Procedures	
a.	Squeeze the two locktabs simultaneously and push in the purge valve barrel.	
b.	Release the locktabs while still pushing on the barrel until the locktabs are engaged.	
c.	Confirm purge valve closing, either by flow changes or by visually confirming the barrel is no longer extended.	

4.8 PURGE VALVE OPERATION

BASIC DATE May 1969   CHANGE DATE June 1971   PAGE 4-56

CSD-A-789-(2)III
APOLLO OPERATIONS HANDBOOK - EMU

STEP	PROCEDURE	REMARKS
4.8.4	Purge Valve Removal Procedures	
a.	Release gas connector lock-lock.	
b.	Lift gas connector locktabs and rotate to release position.	
c.	Remove purge valve from gas connector.	

4.8 PURGE VALVE OPERATION

BASIC DATE May 1969   CHANGE DATE June 1971   PAGE 4-57

CSD-A-789-(2) III
APOLLO OPERATIONS HANDBOOK - EMU

STEP	PROCEDURE	REMARKS
4.9		

BASIC DATE May 1969   CHANGE DATE June 1971   PAGE 4-58

CSD-A-789-(2) III
APOLLO OPERATIONS HANDBOOK - EMU

STEP	PROCEDURE	REMARKS
4.10	EMU INFLIGHT MAINTENANCE	
4.10.1	PGA and LEVA Inflight Examination and Maintenance	This section contains procedures for examining, cleaning, lubricating, and repairing of PGA and LEVA components during flight.
4.10.1.1	PGA and LEVA Inflight Examination	
	During a mission, the PGA and LEVA should be examined for wear and possible damage. A detailed examination should not be attempted unless damage to a component is suspected. If damage is obvious, a more detailed examination and analysis should be performed.	
	The PGA and LEVA should be inspected for the following:	
	a. Loose or broken stitches	
	b. Rips, snags, and abraded areas	
	c. Sharp edges and scratches	
	d. Damaged seals or O-rings	
	e. Proper position and security of components	
	f. Lack of lubrication	
	g. Cleanliness	
	Inflight repairs on items found to be discrepant are possible in certain instances, dependent upon the provisions of the EMU maintenance kit.	

4.10 EMU INFLIGHT MAINTENANCE

BASIC DATE May 1969   CHANGE DATE June 1971   PAGE 4-59

CSD-A-789-(2) III
APOLLO OPERATIONS HANDBOOK-EMU

STEP	PROCEDURE	REMARKS
4.10.1.2	PGA and LEVA Inflight Maintenance	
a.	Cleaning and antifog treatment of pressure helmet, helmet shield, and LEVA viewing areas.	
1.	Cleaning of pressure helmet and helmet shield viewing areas (finger prints, smears, etc. and not lunar dust).	
	CAUTION	
	This procedure should not be used on LEVA as coatings may become damaged.	
(a)	Cut antifog pad container open and extract pad.	
(b)	Apply heaviest possible film of solution, without causing runs, to all inside and outside viewing areas, using a continuous straight line motion.	
(c)	Immediately wipe clean, and buff with drying towel from EMU maintenance kit until visibly clear.	
2.	Cleaning of dust particles from the LEVA and pressure helmet	
(a)	Cleaning lunar dust from LEVA in a pressurized cabin.	
(1)	Blow dust off LEVA.	An alternate step (1) would be to brush LEVA with camera lens brush.
(2)	Pat LEVA softly (<u>DO NOT WIPE</u>) with a wet drying towel.	

4.10 EMU INFLIGHT MAINTENANCE

BASIC DATE May 1969   CHANGE DATE June 1971   PAGE 4-60

CSD-A-789-(2) III
APOLLO OPERATIONS HANDBOOK - EMU

STEP	PROCEDURE	REMARKS
	(3) Wipe LEVA softly with a clean, damp drying towel.	
	(b) Cleaning lunar dust from LEVA while EVA	
	NOTE: If the crewman's vision is obscured by lunar dust, raising the visor to provide visibility within the protection of the sun shade should be attempted first.	
	Wipe LEVA softly with a glove gauntlet, beta bag, flag, etc.	
	(c) Cleaning lunar dust from pressure helmet in a pressurized cabin.	
	(1) Blow dust off pressure helmet	
	(2) Pat pressure helmet softly with an antifog pad	
	(3) Pat pressure helmet until dry with a drying towel.	
	3. Antifog treatment of pressure helmet	
	(a) Cut antifog pad container open and extract pad.	
	(b) Apply heaviest possible film of solution, without causing runs, on all inside viewing areas using a continuous straight line motion.	
	(c) Immediately wipe dry and buff with drying towel.	

4.10 EMU INFLIGHT MAINTENANCE

BASIC DATE May 1969   CHANGE DATE June 1971   PAGE 4-61

CSD-A-789-(2) III
APOLLO OPERATIONS HANDBOOK - EMU

STEP	PROCEDURE	REMARKS
	NOTE: If the pressure helmet cleaning procedure (a1) has just been completed, steps 3b and 3c may be omitted.	
	(d) Apply second coat as in step 3b using clean side of antifog pad.	Coat one helmet with one pad.
	(e) Immediately wipe dry and buff with a new drying towel until visibly clear.	
b.	Maintenance of seals and O-rings	All accessible seals and O-rings may be lubricated in flight.
1.	Removal of seal or O-ring	
	(a) Fit the contoured end of the seal removal tool between the seal O-ring and seat.	
	(b) Rotate the tool circumferentially around until the seal O-ring is free of the recess, and remove tool and O-ring.	
2.	Inspection of removed seal or O-ring	
	(a) Inspect removed seal and O-ring for cuts, abrasions, or breaks in surface as well as irregularities in shape.	
	(b) If seal and O-ring are not faulty, lubricate and install. Replace if O-ring is faulty.	
3.	Lubrication of seal and O-ring	
	(a) Obtain lubrication pad from maintenance kit.	

4.10 EMU INFLIGHT MAINTENANCE

BASIC DATE May 1969   CHANGE DATE June 1971   PAGE 4-62

CSD-A-789-(2) III
APOLLO OPERATIONS HANDBOOK - EMU

STEP	PROCEDURE	REMARKS
	(b) Wipe seal and O-ring with pad, being careful not to get lubricant on any other part of PGA.	
	4. Installation of seal and O-ring.	
	(a) Cut the pouch in the maintenance kit to remove replacement seal and O-ring.	
	CAUTION	
	Use care to avoid cutting the seal and O-ring.	
	(b) Remove replacement and lubricate.	
	(c) Install seal and O-ring into opening.	The seal removal tool can be used to facilitate installation of seals.
	c. Bladder Repair	
	Small punctures in the bladder portion of the PGA may be repaired in flight provided the structural integrity of the PGA is not greatly impaired.	Punctures of sufficient magnitude to degrade the restraint quality of the glove bladder may be repaired by a patch. However, the glove will not be used but retained for emergency use.
	1. Determine location of leakage and obtain a repair patch from maintenance kit.	
	2. Cut repair patch to desired size. The repair patch shall not extend more than one-fourth inch beyond the damaged area.	

4.10 EMU INFLIGHT MAINTENANCE

CSD-A-789-(2) III
APOLLO OPERATIONS HANDBOOK - EMU

STEP	PROCEDURE	REMARKS
3.	Remove backing from patch and place adhesive side of patch over damaged area. The patch shall be applied to inside of PGA.	
4.	Apply pressure to ensure positive bond.	
4.10.2	Bioinstrumentation Inflight Repair	
a.	Replacement of loose electrode	
1.	Remove all trace of old electrode paste from electrode site.	
2.	Replace existing electrode using paste P/N SEB42100014 and electrode attachment assembly P/N SEB42150035.	Located in medical accessories kit.
3.	Cover electrode with micropore covering P/N SB-AE-005408.	
b.	Replacement of electrode harness	
1.	Obtain spare electrode harness, and attach each electrode as described in step 4.8.2-a.	Located in medical accessories kit.
2.	Attach electrode harness to signal conditioners. The connectors should be finger tight.	
	CAUTION	
	Do not overtighten connectors.	

4.10 EMU INFLIGHT MAINTENANCE

BASIC DATE May 1969   CHANGE DATE June 1971   PAGE 4-64

CSD-A-789-(2) III
APOLLO OPERATIONS HANDBOOK - EMU

STEP	PROCEDURE	REMARKS
c.	Permanent removal of bioinstrumentation system	
	Remove bioinstrumentation system and cover exposed end of the PGA electrical umbilical and/or T-adapter cable with tape P/N SEB12100050-201 (on board).	

4.10 EMU INFLIGHT MAINTENANCE

CSD-A-789-(2) III
APOLLO OPERATIONS HANDBOOK - EMU

STEP	PROCEDURE	REMARKS
4.11	MISCELLANEOUS PROCEDURES	
4.11.1	PLSS Gas Trap Activation	
a.	EV activation (by other crewman)	
1.	Shift PLSS to extreme left.	
2.	Open gas trap guard.	
3.	Depress gas trap button for 5 seconds then release.	Cooling should be improved in 3 minutes.
4.	Close guard.	
5.	Realine PLSS.	
b.	Pressurized cabin activation with PLSS doffed	
1.	Connect PLSS multiple water connector to suit connector.	
2.	Switch PLSS pump to ON.	
3.	Cycle H_2O diverter valve slowly (three times).	
4.	Switch PLSS pump to OFF.	
5.	Disconnect multiple water connector from suit.	
6.	Connect LM water supply hose to PLSS fill connector and open supply valve.	
7.	Open gas trap guard.	

4.11 MISCELLANEOUS PROCEDURES:

BASIC DATE May 1969   CHANGE DATE June 1971   PAGE 4-66

CSD-A-789-(2)III
APOLLO OPERATIONS HANDBOOK - EMU

STEP	PROCEDURE	REMARKS
8.	Depress gas trap button until water is observed at the vent on top of the gas trap; then release.	
9.	Close guard.	
10.	Close LM water supply valve and disconnect supply hose.	
11.	Replace fill connector cap.	
12.	Close PLSS recharge access door.	
4.11.2	<u>Camera Mounting on RCU</u>	
a.	Outside LM beginning of EVA	
1.	Crewman receives camera with mounting bracket attached.	
2.	Crewman will center the camera bracket (female) at the front center of the RCU and mate the two brackets (camera and RCU halves).	
3.	Push camera and bracket down until lock is in place.	
b.	Release of camera and bracket (assumes crewman unassisted)	
1.	Place right hand under camera and bracket and apply a small force upward.	
2.	Place left thumb or forefinger on tab release lever on front of RCU.	

4.11 MISCELLANEOUS PROCEDURES

BASIC DATE May 1969   CHANGE DATE June 1971   PAGE 4-67

CSD-A-789-(2) III
APOLLO OPERATIONS HANDBOOK-EMU

STEP	PROCEDURE	REMARKS
3.	Push release lever to the right while applying upward force from base of camera and lift camera from RCU mounting.	
4.11.3	**BSLSS Bag to PLSS Donning**	
a.	Insert BSLSS tether pins into PLSS upper hard point mounts (bag is oriented so that pip pin on end of longer strap is in mounting hole near auxiliary feedwater tank).	a. One BSLSS for two crewmen. b. Assistance of second crewman required to don BSLSS. c. Both crewmen EV on lunar surface.
b.	Depress head of tether pins and complete insertion.	
c.	Attach doffing tethers as desired.	
4.11.4	**BSLSS Bag to PLSS Doffing**	
a.	Grasp end of doffing tethers.	
b.	Extend tethers until tether pins are released.	
c.	Discard or stow BSLSS as applicable.	
4.11.5	**BSLSS Stowage on LRV**	BSLSS is stowed on LRV during lunar traverse and between EVA's.
	Hang the BSLSS on the back of the LMP seat on the LRV. This is done by looping the Velcro strap attached to the seat back through the BSLSS bag strap and mating with the Velcro on the front of the seat back.	

4.11 MISCELLANEOUS PROCEDURES

BASIC DATE May 1969   CHANGE DATE June 1971   PAGE 4-68

APOLLO OPERATIONS HANDBOOK - EMU

STEP	PROCEDURE	REMARKS
4.11.6	**BSLSS Donning and Activation Procedure**	
a.	Remove BSLSS assembly from stowage.	
b.	Attach tether (adjacent to flow divider) to the left PGA LM restraint ring of crewman which has operational PLSS.	Assumption: Operational PLSS on the right side of the non-operational PLSS.
c.	Remove dust cover from H_2O flow divider.	
d.	Disconnect PLSS H_2O connector from the PGA with operating PLSS.	The following guides should be followed when attaching either of the BSLSS or PLSS water connectors to the PGA, or the PLSS water connector to the BSLSS H_2O flow divider.
e.	Connect BSLSS H_2O flow divider to the PGA with operational PLSS.	
f.	Turn off pump of failed PLSS.	Attach H_2O connectors on buddy basis:
g.	Disconnect PLSS H_2O connector from the PGA with failed PLSS, and secure.	a. Buddy crewman insert male connector into vacant female connector on PGA or BSLSS flow divider.
h.	Disconnect BSLSS H_2O connector from H_2O flow divider.	b. Buddy place one arm and hand behind assisted crewman for stabilization.
i.	Attach other tether to the right LM restraint D-ring of the crewman with nonoperational PLSS.	c. Press male connector into female connector until lock ring is actuated.
j.	Connect BSLSS H_2O connector to the PGA with failed PLSS.	
k.	Connect operational PLSS water connector to H_2O flow divider.	
4.11.7	**BSLSS Disconnect Procedure**	
a.	Disconnect BSLSS water connector from PGA with failed PLSS.	

4.11 MISCELLANEOUS PROCEDURES

CSD-A-789-(2) III
APOLLO OPERATIONS HANDBOOK - EMU

STEP	PROCEDURE	REMARKS
b.	Disconnect tether from each crewman's PGA LM restraint attachment.	
c.	Crewman with failed PLSS ready to ingress LM.	
d.	Disconnect PLSS water connector from BSLSS water flow divider.	
e.	Disconnect BSLSS water flow divider connector from PGA.	
f.	Discard BSLSS.	
g.	Connect operational PLSS water connector to PGA.	
h.	Crewman with operational PLSS ready to ingress LM.	
4.11.8	<u>Donning and Doffing Helmet Shield</u>	
4.11.8.1	Donning Helmet Shield	
	Expand shield slip hole over helmet feedport, rotate opposite side over helmet, and allow to contract into place.	
4.11.8.2	Doffing Helmet Shield	
	Expand shield, rotate about feedport, and disengage from feedport.	
4.11.9	<u>Donning and Doffing Neck Dam</u>	
4.11.9.1	Donning Neck Dam	The neck dam is donned before water egress to prevent water from entering the open neck area of the TLSA.

4.11 MISCELLANEOUS PROCEDURES

BASIC DATE May 1969   CHANGE DATE June 1971   PAGE 4-70

CSD-A-789-(2) III
APOLLO OPERATIONS HANDBOOK - EMU

STEP	PROCEDURE	REMARKS
a.	Place helmet-attaching ring lock assembly in ENGAGE position.	
b.	Slip neck dam over head so that eyelet tab is forward.	
c.	Aline neck dam ring eyelet tab with index marks on helmet-attaching ring and snap tab in place.	
d.	Continue pushing other tabs down into place until all tabs of neck dam are latched.	
e.	Place the lock assembly into LOCK.	
4.11.9.2	Doffing Neck Dam	
a.	Unlock helmet-attaching ring and separate neck-dam.	
b.	Pull neck dam over head.	The neck dam is used only for water egress.
4.11.10	Insuit Drinking Device (ISDD) Installation and Use	
4.11.10.1	ISDD Installation	
a.	Remove the ISDD spacecraft overwrap by cutting with scissors along one edge of the wrapper just inside any of the heat-sealed seams. Remove the device and discard the overwrap.	Preferred method is to install ISDD prior to donning PGA. If gas is entrapped within the ISDD, actuate the mouthpiece tilt valve, gently squeeze the bag to expel the entrapped gas, and then allow the mouthpiece tilt valve to return to the closed position. This procedure may also be performed following 4.11.10.2.C, if required.
b.	Remove the PGA liner from its Velcro and snap attachments in the frontal area of the PGA neck ring.	

BASIC DATE May 1969 CHANGE DATE June 1971 PAGE 4-71

CSD-A-789-(2) III
APOLLO OPERATIONS HANDBOOK - EMU

STEP	PROCEDURE	REMARKS
c.	Insert the ISDD (mouthpiece to crewman's left) into the PGA neck opening between the liner and suit bladder, positioning the mouthpiece as desired by crewman preference. Once positioned, press the Velcro hook of the ISDD onto the Velcro pile on the suit bladder.	
d.	Replace the liner into position, and attach the Velcro hook on the liner to the Velcro pile on the Velcro pile on the ISDD. Replace any liner snaps removed in step b that are not covered by the Velcro of the ISDD.	
e.	Proceed with PGA donning.	
4.11.10.2	ISDD Filling	
a.	Insert the water dispenser into the ISDD fill valve.	The fill valve of the ISDD is identical to that utilized on the Apollo rehydratable food packages and operates in the same manner.
b.	Depress the trigger of the water dispenser and tilt the ISDD mouthpiece tilt valve. Continue to fill the device until water venting is noted at the mouthpiece.	
c.	Release the water dispenser trigger, allow the mouthpiece tilt valve to return to the closed position, and remove the water dispenser from the ISDD fill valve.	The following alternate sequence may be utilized for ground operations and in the LM on the lunar surface between EVA periods. With the ISDD installed in the PGA, fill ISDD to the label fill line expelling entrapped gas through the mouthpiece tilt valve and then don the PGA.
d.	Proceed with pre-EVA preparations and donning of PGA helmet and gloves.	

4.11 MISCELLANEOUS PROCEDURES

CSD-A-789-(2) III
APOLLO OPERATIONS HANDBOOK - EMU

STEP	PROCEDURE	REMARKS
4.11.10.3	ISDD Drinking	
a.	Turn head to the left and grasp the ISDD mouthpiece in the mouth.	
b.	Activate the ISDD mouthpiece tilt valve by bending the mouthpiece tube, and suck water up the tube as if using a straw.	
c.	After obtaining sufficient water, release grasp on the device mouthpiece.	The ISDD may be refilled as required in accordance with section 4.11.10.2 prior to each subsequent EVA period.
4.11.10.4	ISDD Doffing	
a.	Remove the PGA liner from its Velcro/snap attachments and the ISDD Velcro in the frontal area of the PGA neck opening.	
b.	Grasp the ISDD by one end of the Velcro, and remove the device from the PGA.	
c.	Replace the PGA liner to its mating Velcro and snaps.	

4.11 MISCELLANEOUS PROCEDURES

BASIC DATE May 1969   CHANGE DATE June 1971   PAGE 4-73

STEP	PROCEDURE	REMARKS
4.11.11	OPS Oxygen Usage for Metabolic Makeup — LM and CM	
a.	Verify OPS O$_2$ connector locked in stowage plate.	
b.	Move OPS actuation lever to ON.	Flow rate in this mode is approximately 0.3 lb/hr.
c.	Use as required, and then move OPS actuating lever to OFF.	
4.11.12	OPS Bleed Down — LM and CM	
a.	Verify that OPS actuation lever is OFF.	
b.	Unstow OPS O$_2$ connector from O$_2$ connector stowage plate. **CAUTION** Do not unstow access flaps as they provide hose restraint.	
c.	Move OPS actuation lever to ON.	Flow rate in this mode is approximately 250 lb/hr at 5880 psi.
d.	Use as required, and then move OPS actuation lever to OFF.	

4.11 MISCELLANEOUS PROCEDURES

SECTION 5

EMU MALFUNCTION PROCEDURES

FOR LUNAR SURFACE EVA

5.1 INTRODUCTION

5.1.1 The malfunction procedures encompass the recognition, diagnosis, and corrective action for system malfunctions. In most cases, the crew is alerted to a malfunction condition by indicators and gages. The malfunction analyses do not contain solutions; such solutions are found in mission rules. The procedures in this section cover significant single failures and are not intended to replace the detailed failure modes-effects analyses published in other documents.

5.1.2 The malfunction procedures are for use during Apollo missions where an EMU having an SV 706100-7 PLSS will be worn. The procedures have been classified as (1) emergency, (2) EVA, and (3) post-EVA. (The post-EVA period commences when the LM cabin pressure reaches 3.5 psia.)

5.1.3 For maximum safety, all emergency procedures should be memorized so action can be taken immediately when the malfunction occurs. The EVA procedures do not need to be memorized since they are provided through voice communications except for steps which cover EVA procedures for loss of voice communications.

5.1.4 The post-EVA procedures are designed to extract a maximum amount of information on any observed anomaly since the PLSS/OPS and associated hardware would not normally be returned to earth for postflight analysis. The emergency procedures are devised so that telemetry is not used because telemetry data may not be available to the crewman. (Telemetry should be employed to aid the emergency procedures, however, if it is possible to do so.)

5.1.5 The procedures and remarks are representative of a nominal EMU. Values and quantities, which are characteristic of an individual EMU and which can be established only be testing the actual EMU to be used in the flight, are underlined in each case.

5.2 FORMAT OF MALFUNCTION PROCEDURES

5.2.1 Malfunction procedures are presented in the format of logic-flow block diagrams. Diagram blocks represent procedural steps. Those blocks outlined with double lines and containing capitalized statements indicate system failures. Within blocks, statements preceded by black dots (●) indicate required actions.

5.2.2 The malfunction procedures are presented in three columns headed SYMPTOM, PROCEDURE, and REMARKS. A description and use of each of these columns is as follows:

SYMPTOM The primary purpose of the symptom column is to give a first indication of the malfunction as received by either the crew or telemetry. The possible causes of the malfunction are indicated in this column.

PROCEDURE The procedures column presents a step-by-step logic-flow diagram of actions and decisions used to isolate or correct a malfunction symptom. The remote-event number symbols are used to reference items to the REMARKS column or to refer to other procedural steps.

REMARKS This column will include the following information:

a. Amplifying additional remarks related to the symptom, such as relief valve vents at ____ psid.

b. Amplifying remarks which relate to a decision and/or action items.

c. Explaining resultant system status or operational capability after a failure has been identified.

d. Cautioning or warning, as necessary, to cover conditions that may exist because of a failure

CSD-A-789-(2) III
APOLLO OPERATIONS HANDBOOK - EMU

5.3 MALFUNCTION SYMPTOMS FOR EMU USING PLSS SV 706100-7

Symptom	Page
EMU 1 - Warning tone ON and low vent flow flag "P" (vent-P)	5-5
EMU 2 - Warning tone ON and low PGA pressure flag "O" (PRES-"O")	5-7
EMU 3 - Warning tone ON and high O_2 flow flag "O" (O_2-"O")	5-10
EMU 4 - Warning tone and low feedwater pressure flag "A" (H_2O-"A") while using primary feedwater	5-12
EMU 5 - Warning tone and low feedwater pressure flag "A" (H_2O-"A") while using auxiliary feedwater	5-14
EMU 6 - Warning tone with all warning flags CLEAR	5-16
EMU 7 - PGA pressure gage <3.7 psid and apparently stable and everything else normal	5-18
EMU 8 - PLSS O_2 quantity indicator abnormal reading	5-19
EMU 9 - PGA pressure gage >4.0 psid	5-20
EMU 10 - Loss of pump noise	5-21
EMU 11 - Inadequate cooling of crewman	5-22
EMU 12 - EVA-1 loses voice from MSFN (EVA-1 has voice from EVA-2)	5-24
EMU 13 - EVA-1 loses voice from MSFN (EVA-1 does not receive voice from EVA-2)	5-25
EMU 14 - EVA-1 loses voice from EVA-2 (EVA-1 has voice from MSFN)	5-27
EMU 15 - EVA-2 loses voice from MSFN (EVA-2 receives voice from EVA-1)	5-29
EMU 16 - EVA-2 loses voice from MSFN (EVA-2 does not receive voice from EVA-1)	5-30
EMU 17 - EVA-2 loses voice from EVA-1 (EVA-2 has comm. with MSFN)	5-32
EMU 18 - Loss of voice comm. with EVA-1, EVA-2, or MSFN (two-man EVA) (comm. restoration procedure)	5-34
EMU 19 - EVA loses voice from LM	5-35

CSD-A-789-(2) III
APOLLO OPERATIONS HANDBOOK - EMU

Symptom	Page
EMU 20 - LM loses voice from EVA	5-37
EMU 21 - GT8168P/GT8268P PGA pressure <3.7 psid and apparently steady (no warning tone)	5-39
EMU 22 - GT8168P/GT8268P PGA pressure >4.0 psid	5-40
EMU 23 - GT8182P/GT8282P PLSS O_2 pressure abnormal (no warning tone)	5-41
EMU 24 - GT8110P/GT8210P feedwater pressure <1.2 psia (no warning tone) while using primary feedwater	5-43
EMU 25 - GT8110P/GT8210P feedwater press. <1.2 psia (no warning tone) while using auxiliary feedwater	5-44
EMU 26 - GT8154T/GT8254T LCG H_2O temperature >50° F (with diverter valve in MAX COOLING position and no warning tone)	5-45
EMU 27 - GT8196T/GT8296T LCG H_2O ΔT >11° F (maximum diverter valve position)	5-46
EMU 28 - GT8140C/GT8240C PLSS battery current >3.0 A (no warning tone)	5-47
EMU 29 - GT8140C/GT8240C PLSS battery current <2.3 A (no warning tone)	5-49
EMU 30 - GT8141V/GT8241V PLSS battery voltage <16.0 V dc (no warning tone)	5-51
EMU 31 - GT8170T/GT8270T O_2 temperature <38° F (no warning tone)	5-53
EMU 32 - GT8170T/GT8270T O_2 temperature >50° F and rising (no warning tone)	5-54

BASIC DATE May 1969   CHANGE DATE June 1971   PAGE 5-4

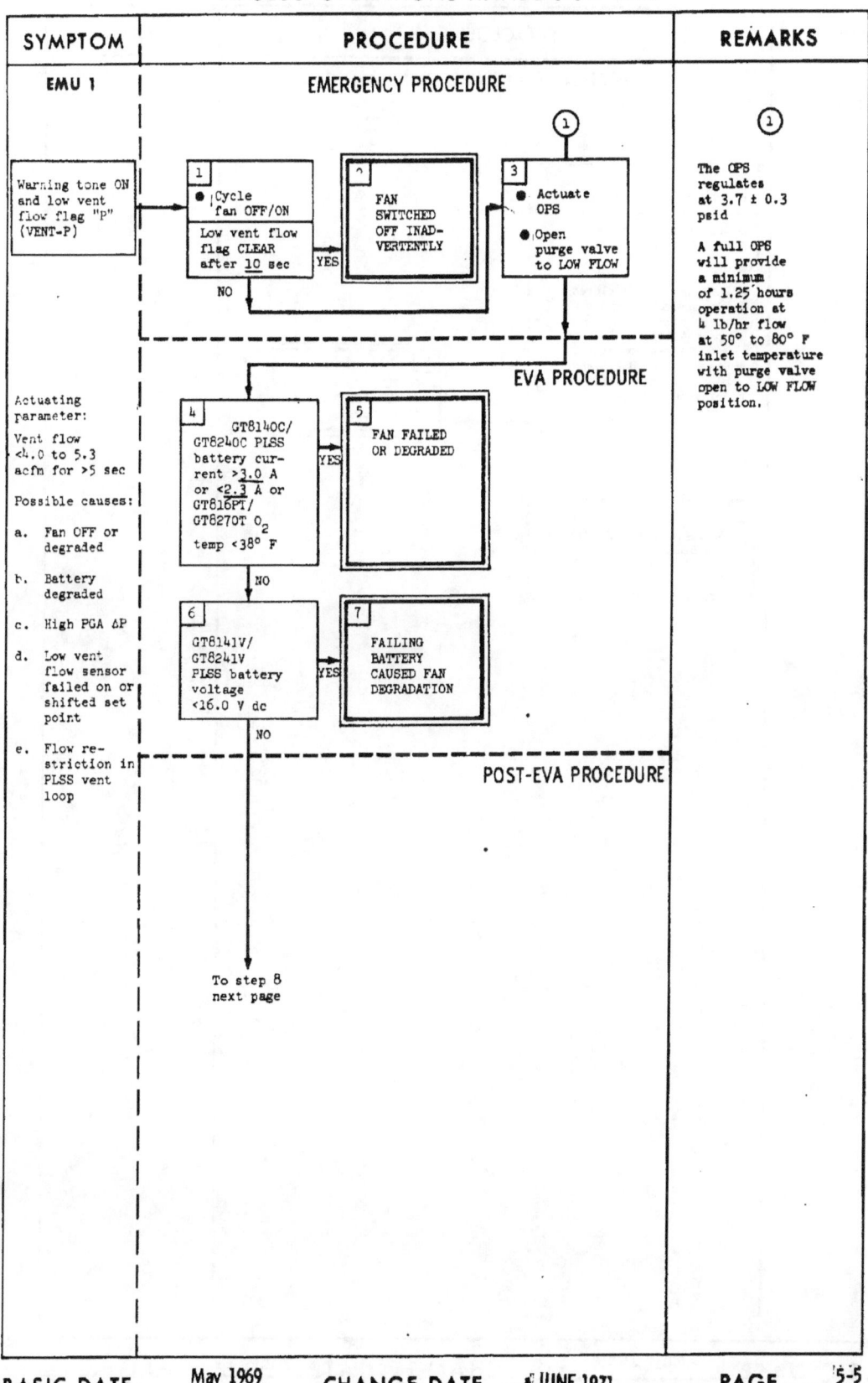

SYMPTOM	PROCEDURE	REMARKS

POST-EVA PROCEDURE (Continued)

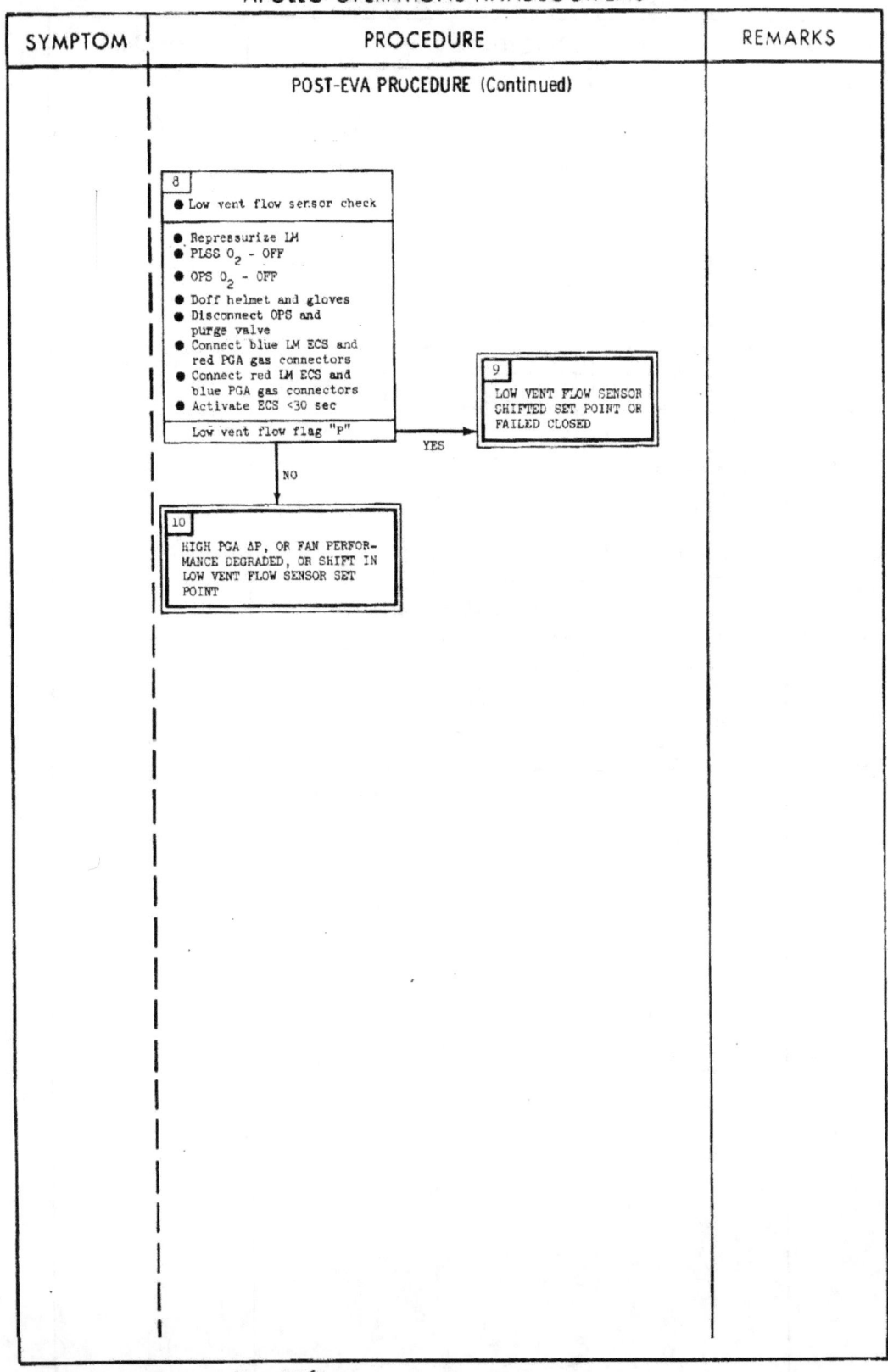

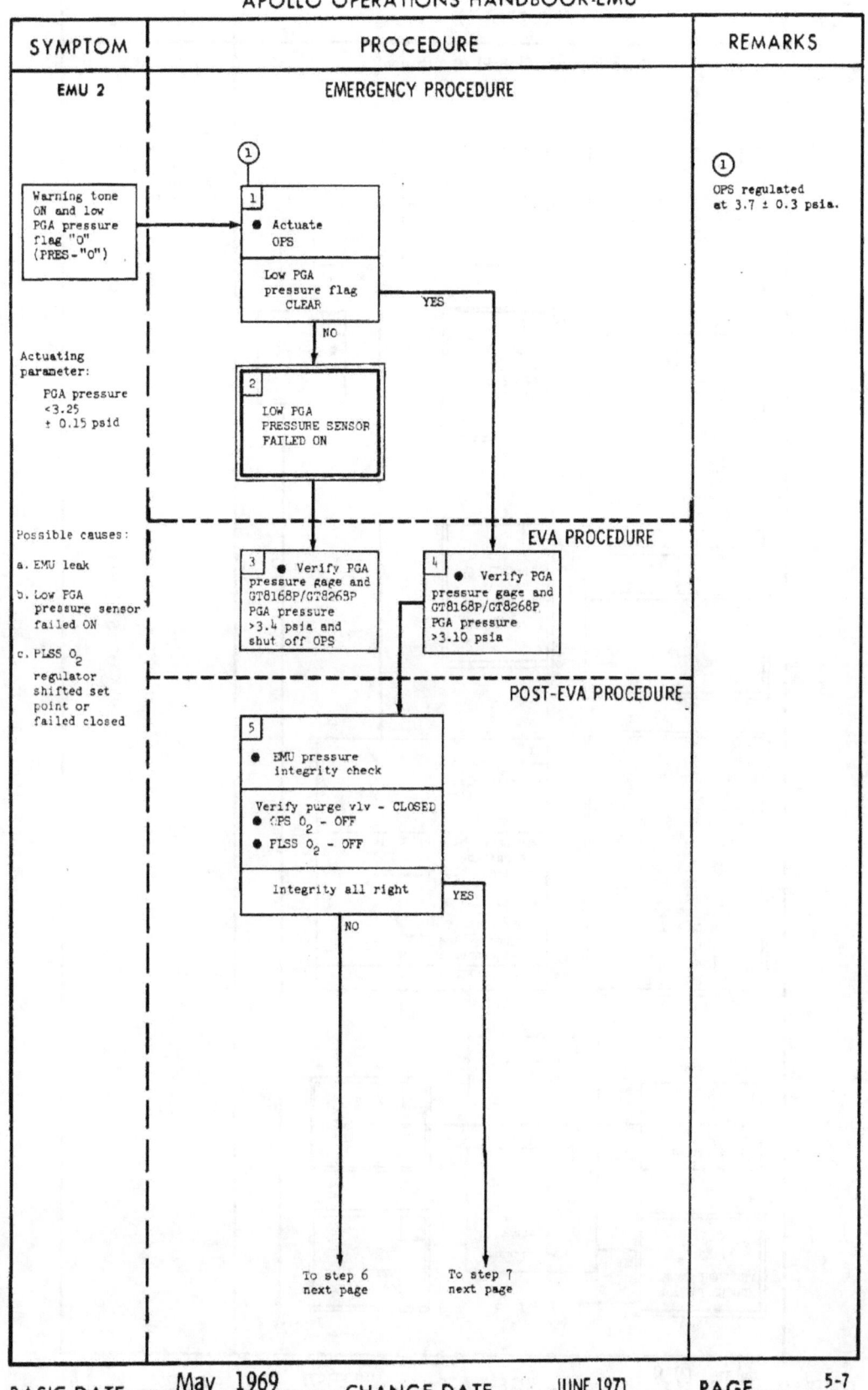

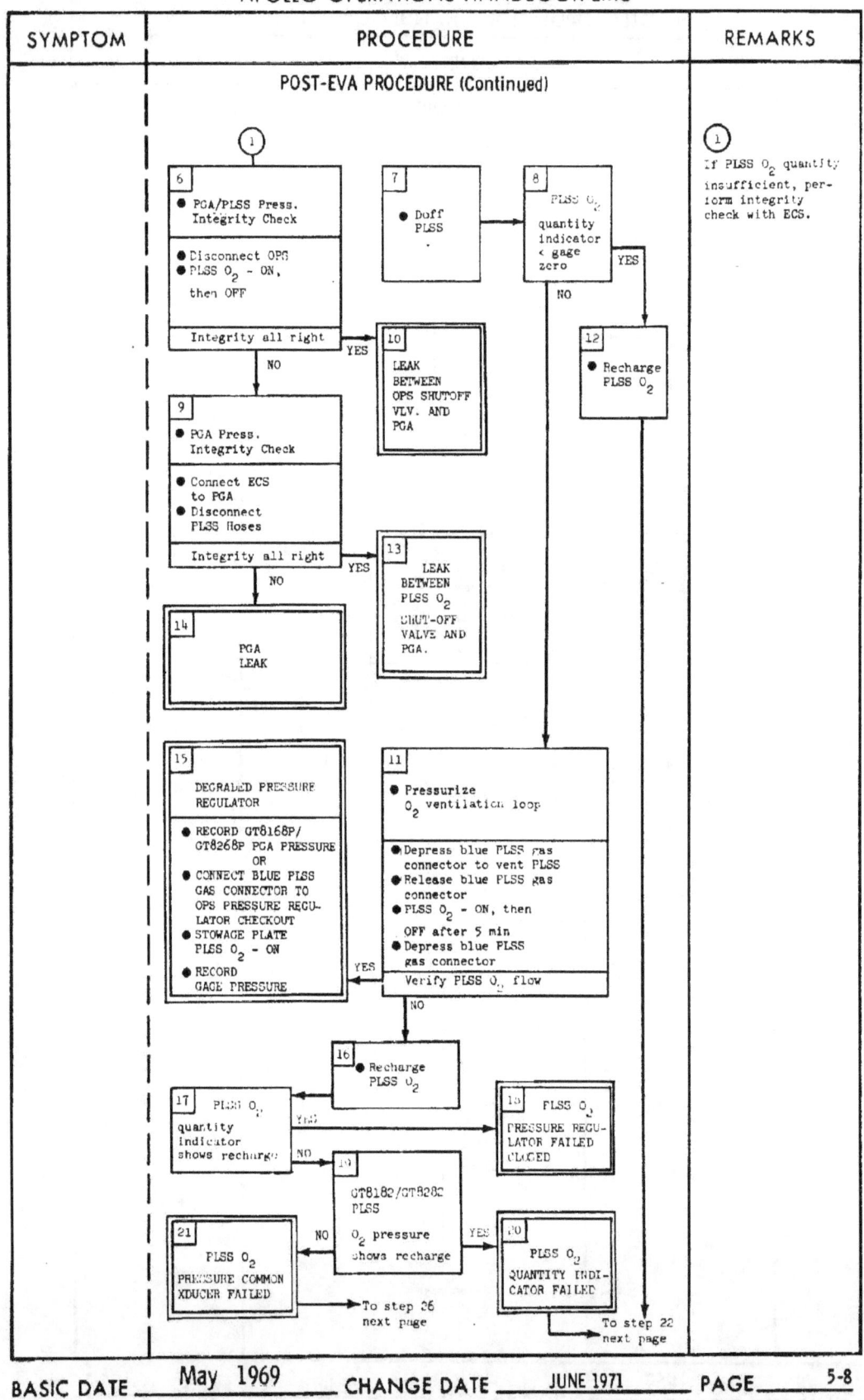

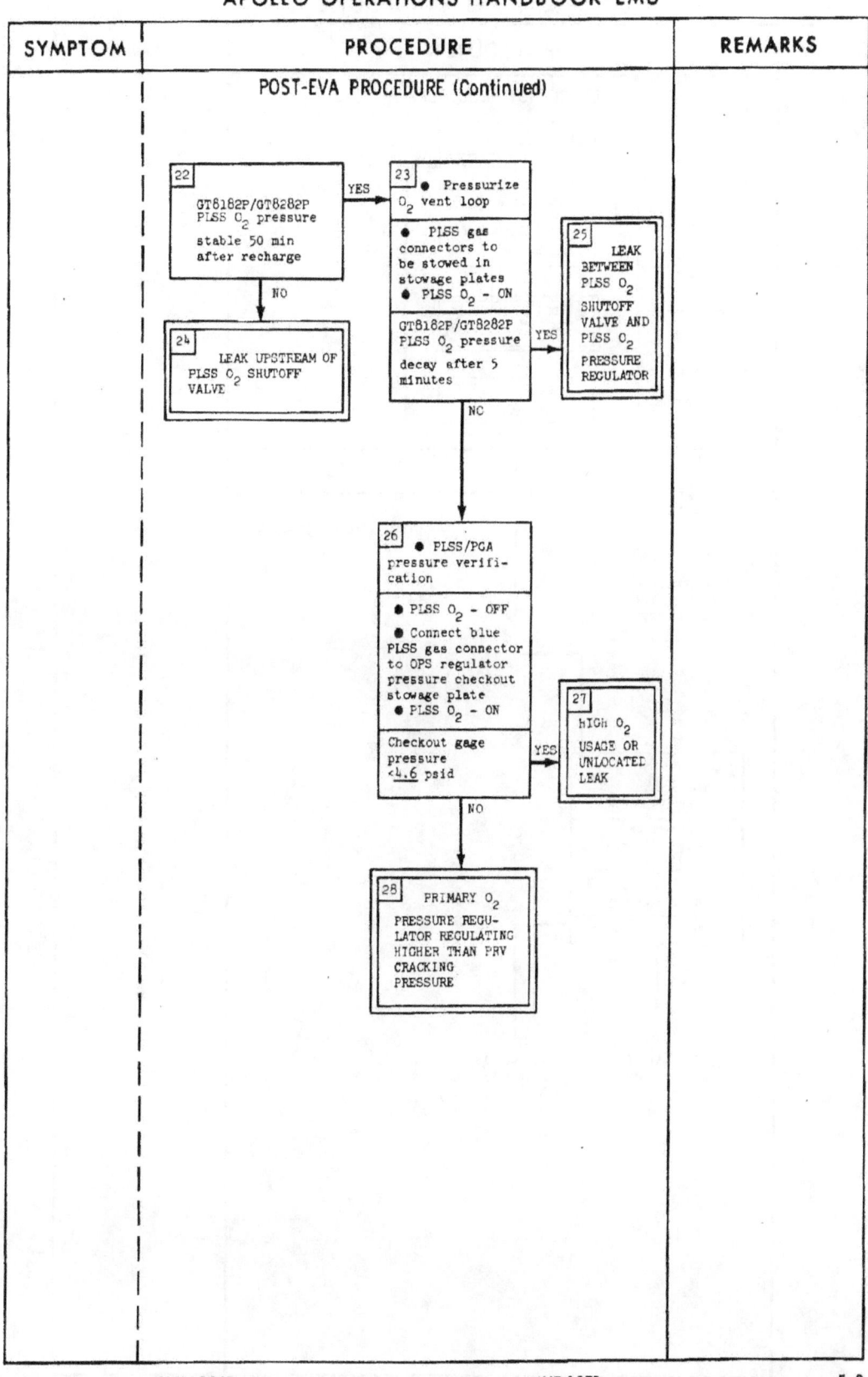

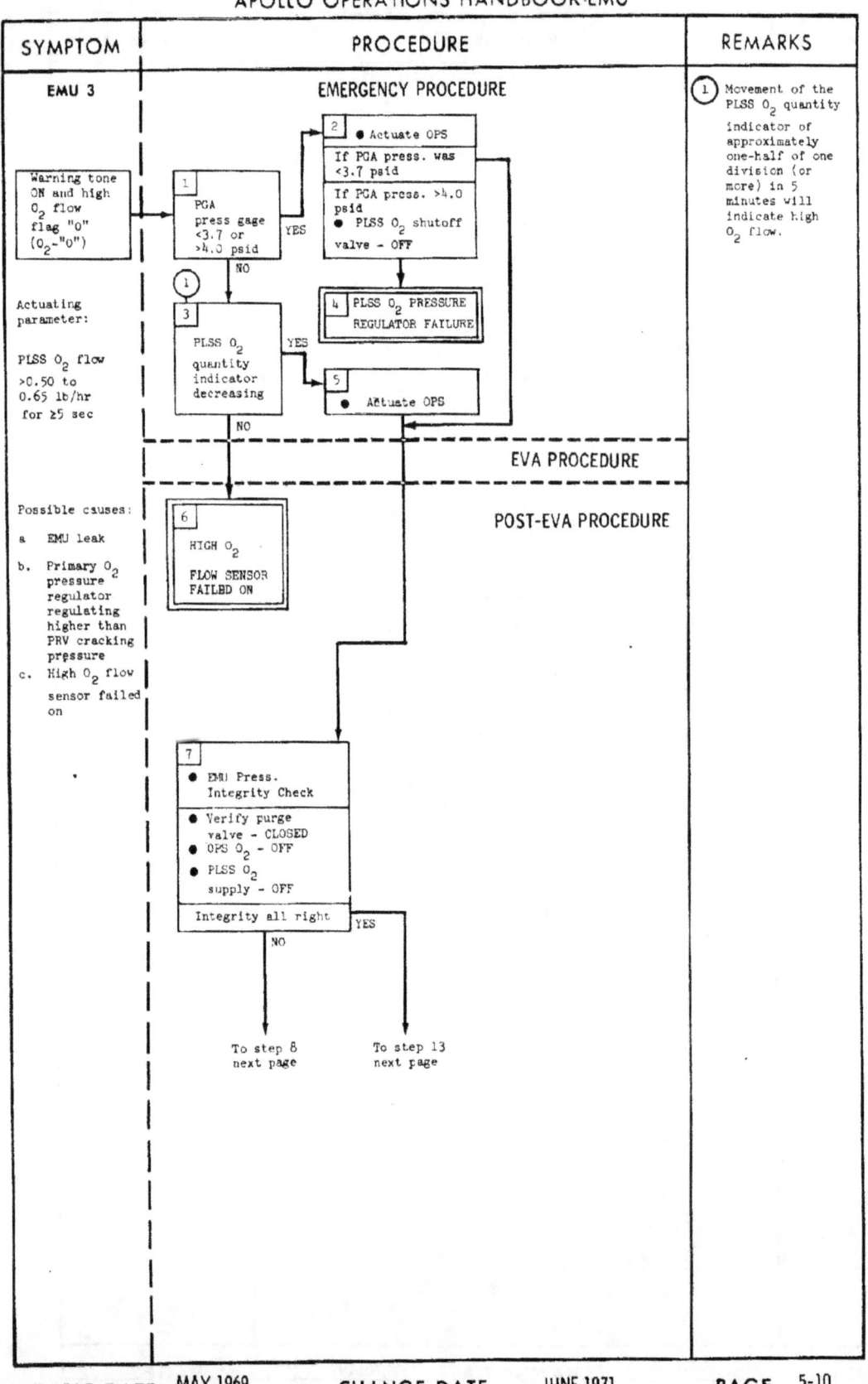

CSD-A-789-(2) III
APOLLO OPERATIONS HANDBOOK-EMU

SYMPTOM	PROCEDURE	REMARKS
	POST-EVA PROCEDURE (Continued) 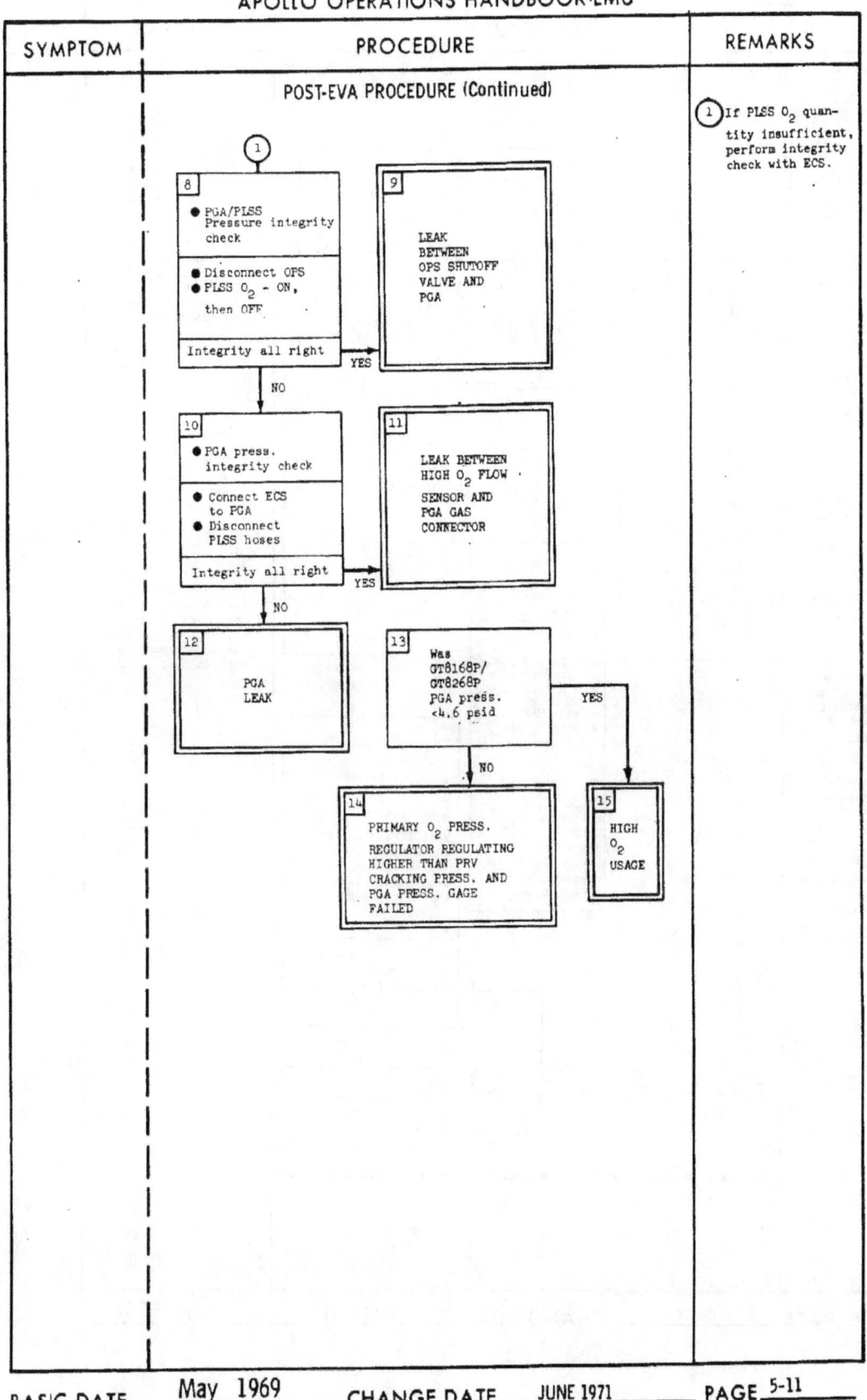	(1) If PLSS O_2 quantity insufficient, perform integrity check with ECS.

BASIC DATE ___ May 1969 ___ CHANGE DATE ___ JUNE 1971 ___

CSD-A-789-(2) III
APOLLO OPERATIONS HANDBOOK-EMU

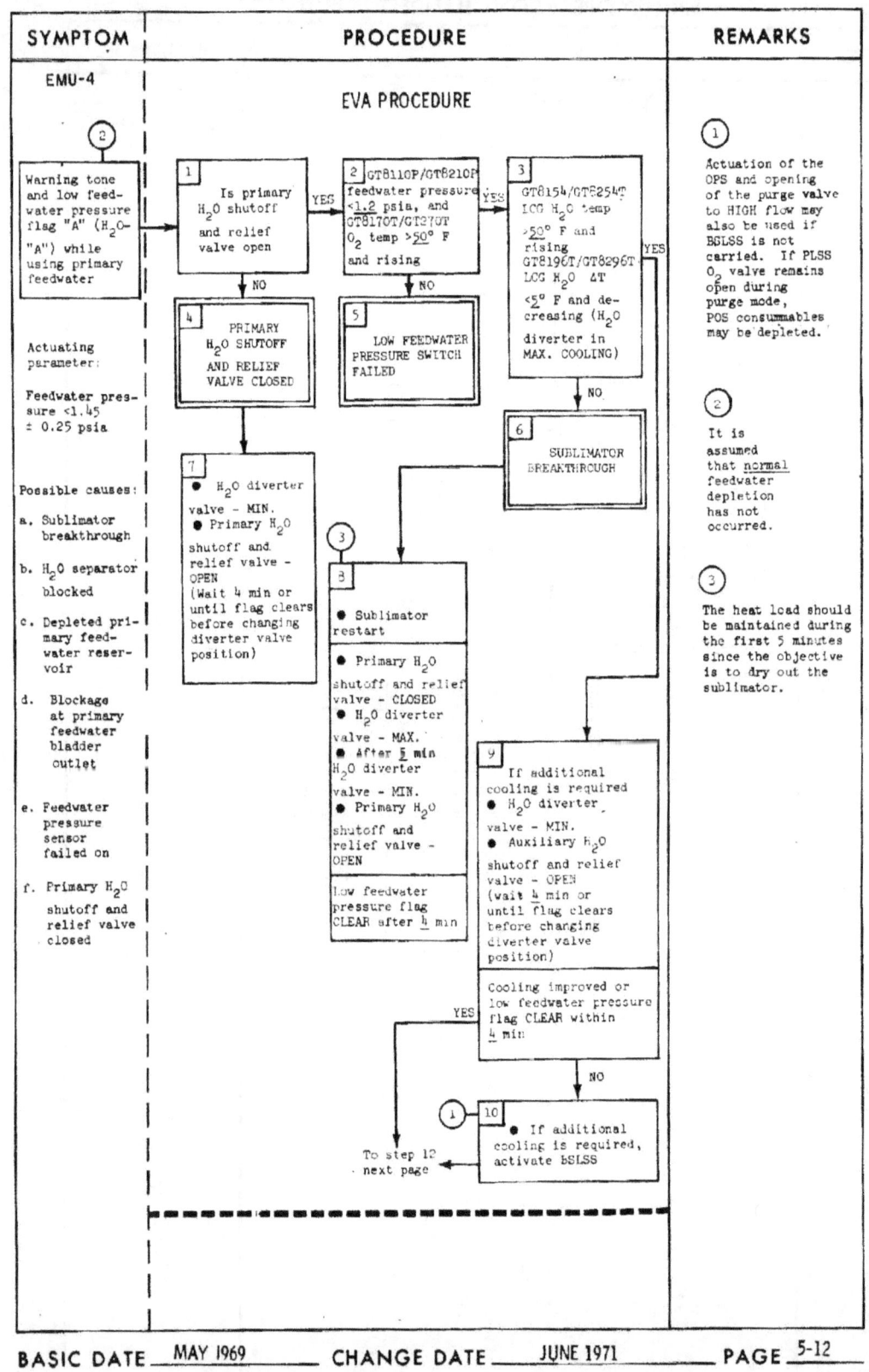

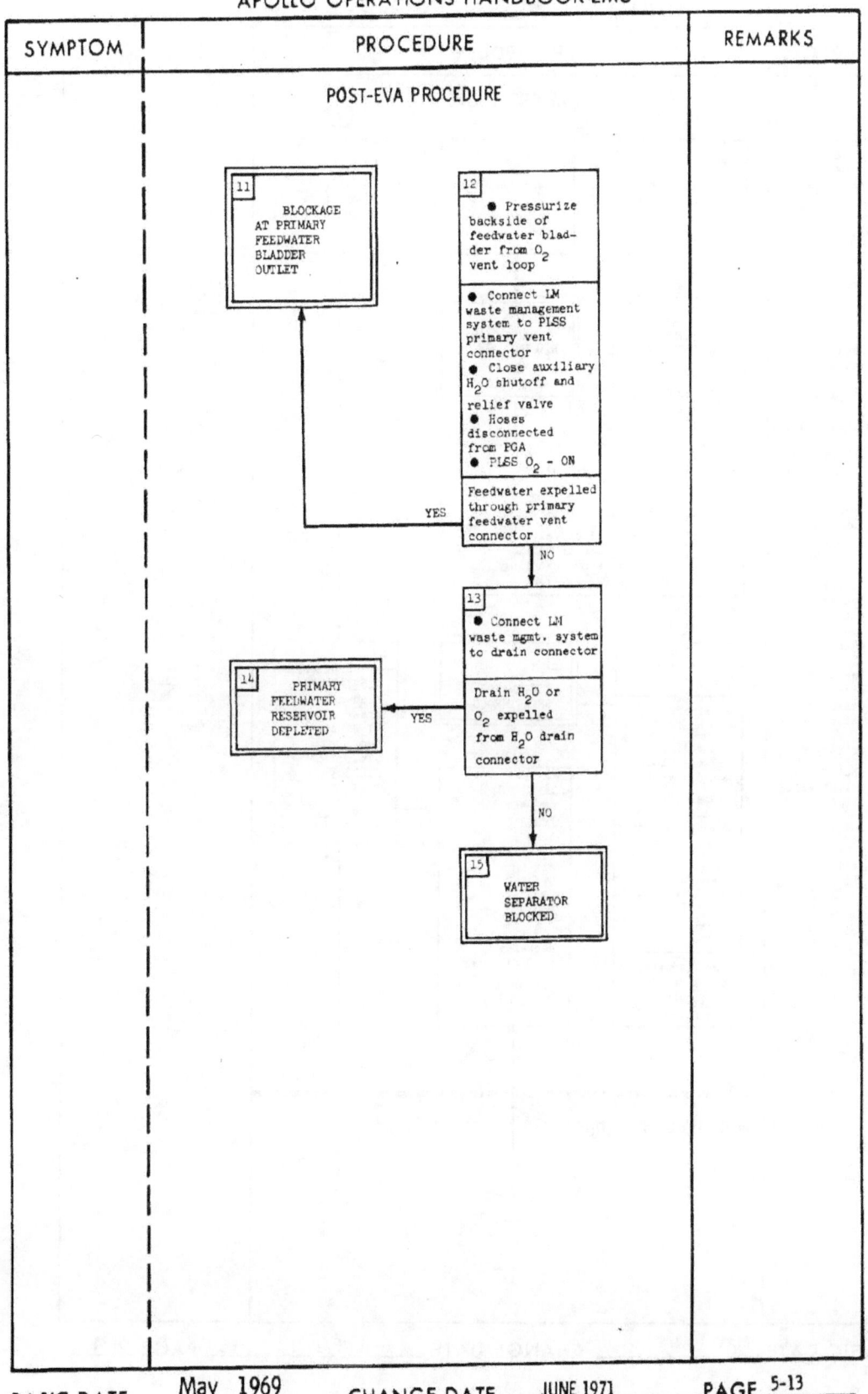

CSD-A-789-(2) III
APOLLO OPERATIONS HANDBOOK - EMU

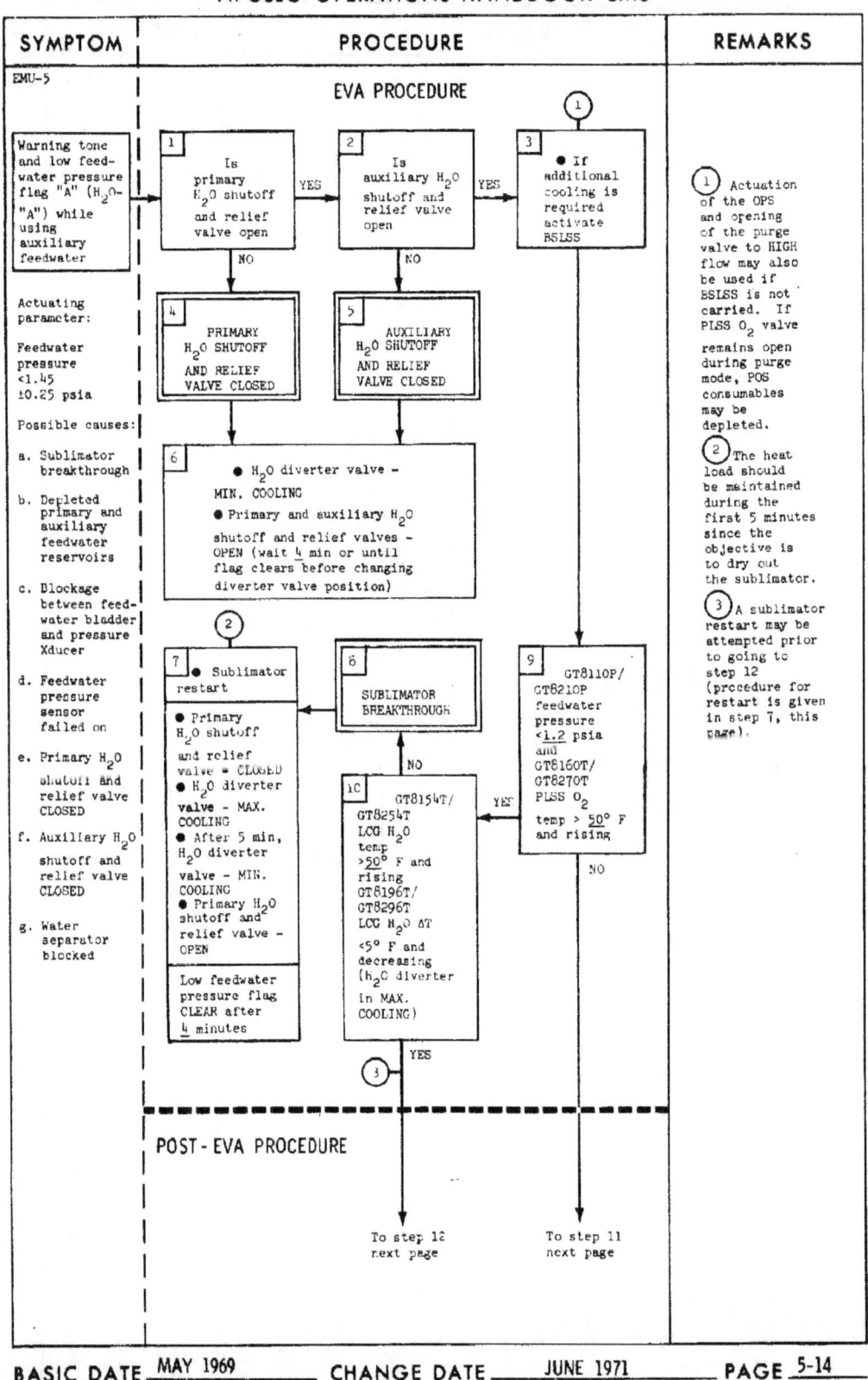

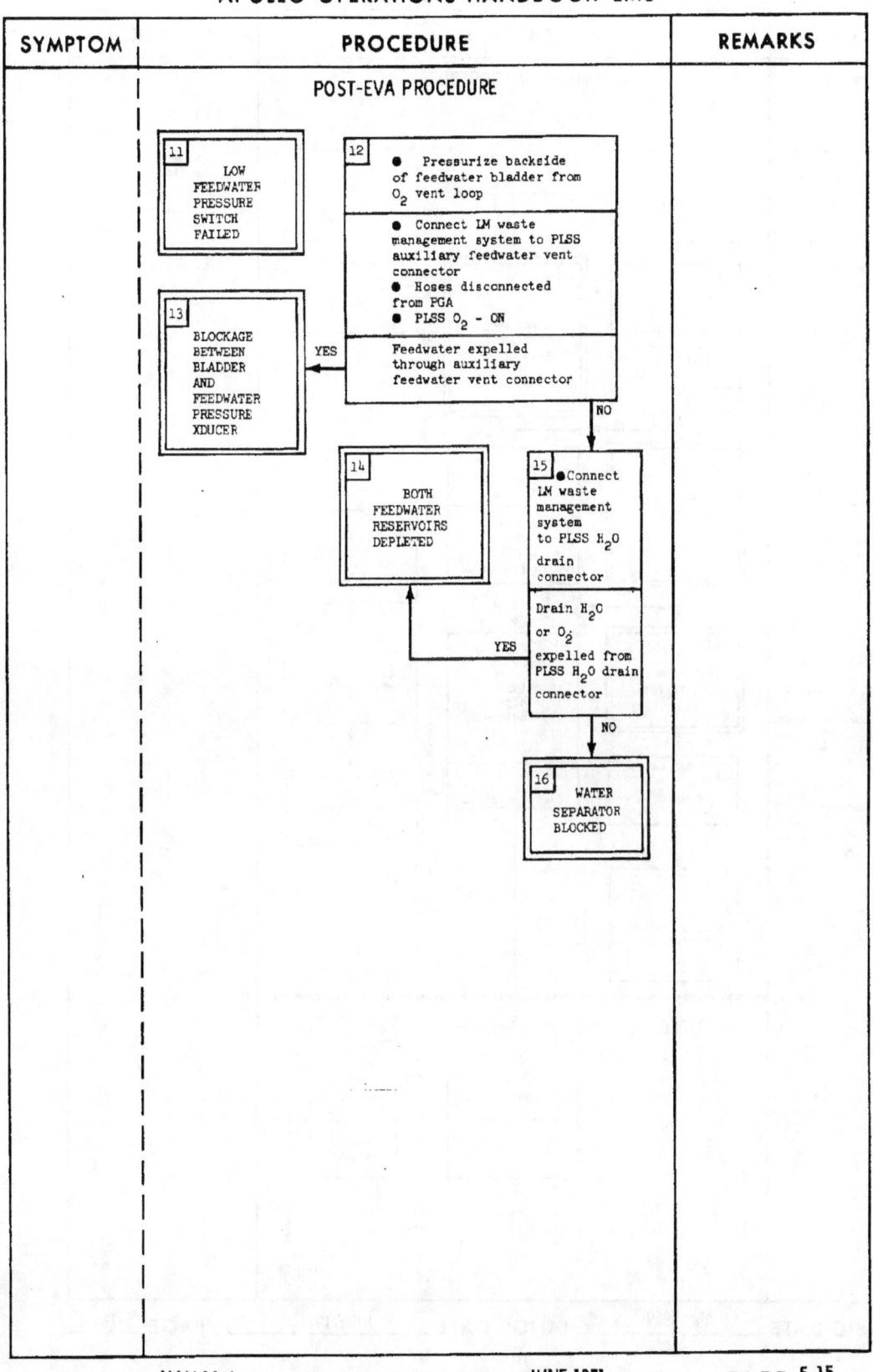

CSD-A-789-(2) III
APOLLO OPERATIONS HANDBOOK - EMU

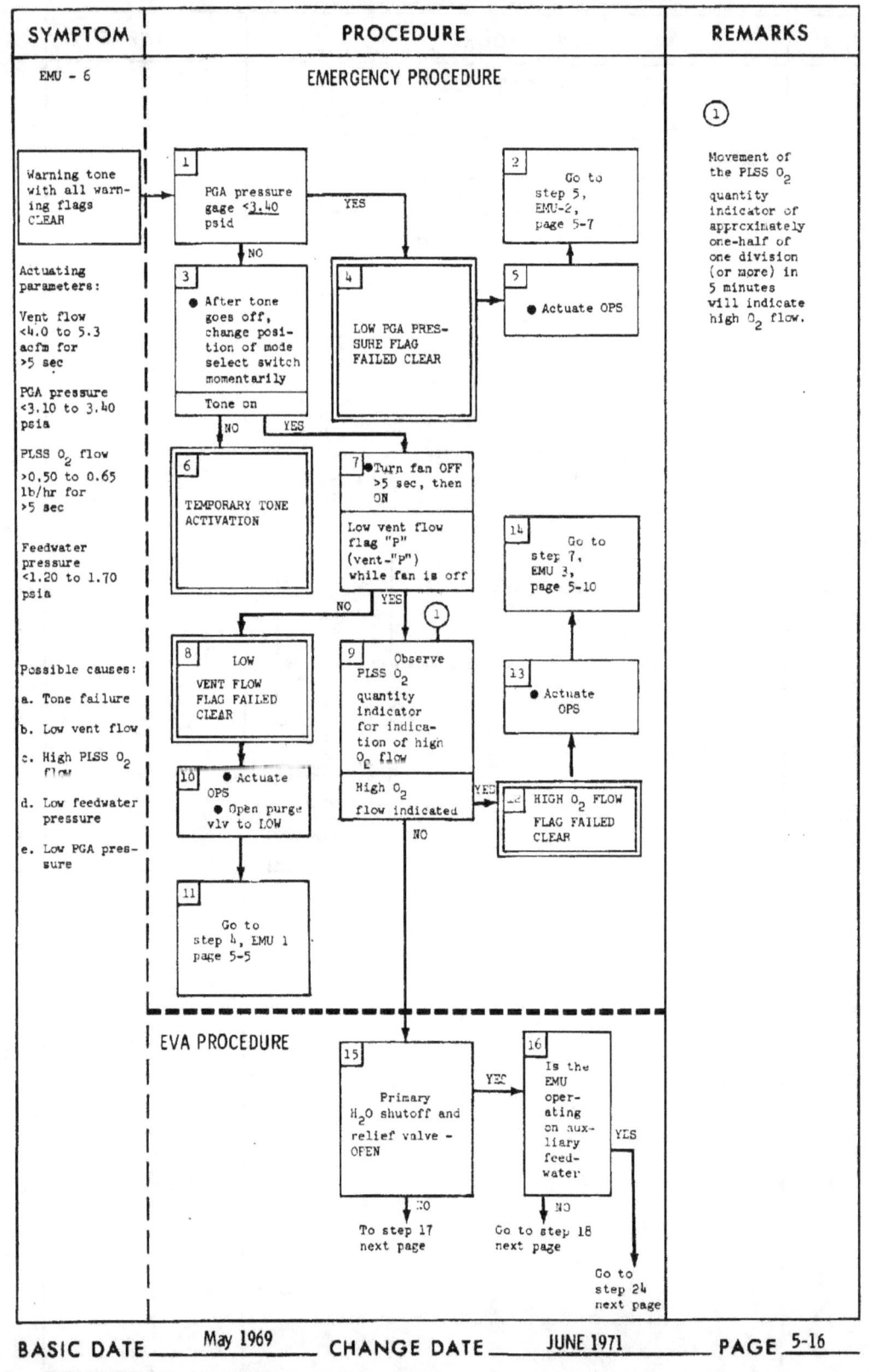

BASIC DATE May 1969   CHANGE DATE JUNE 1971   PAGE 5-16

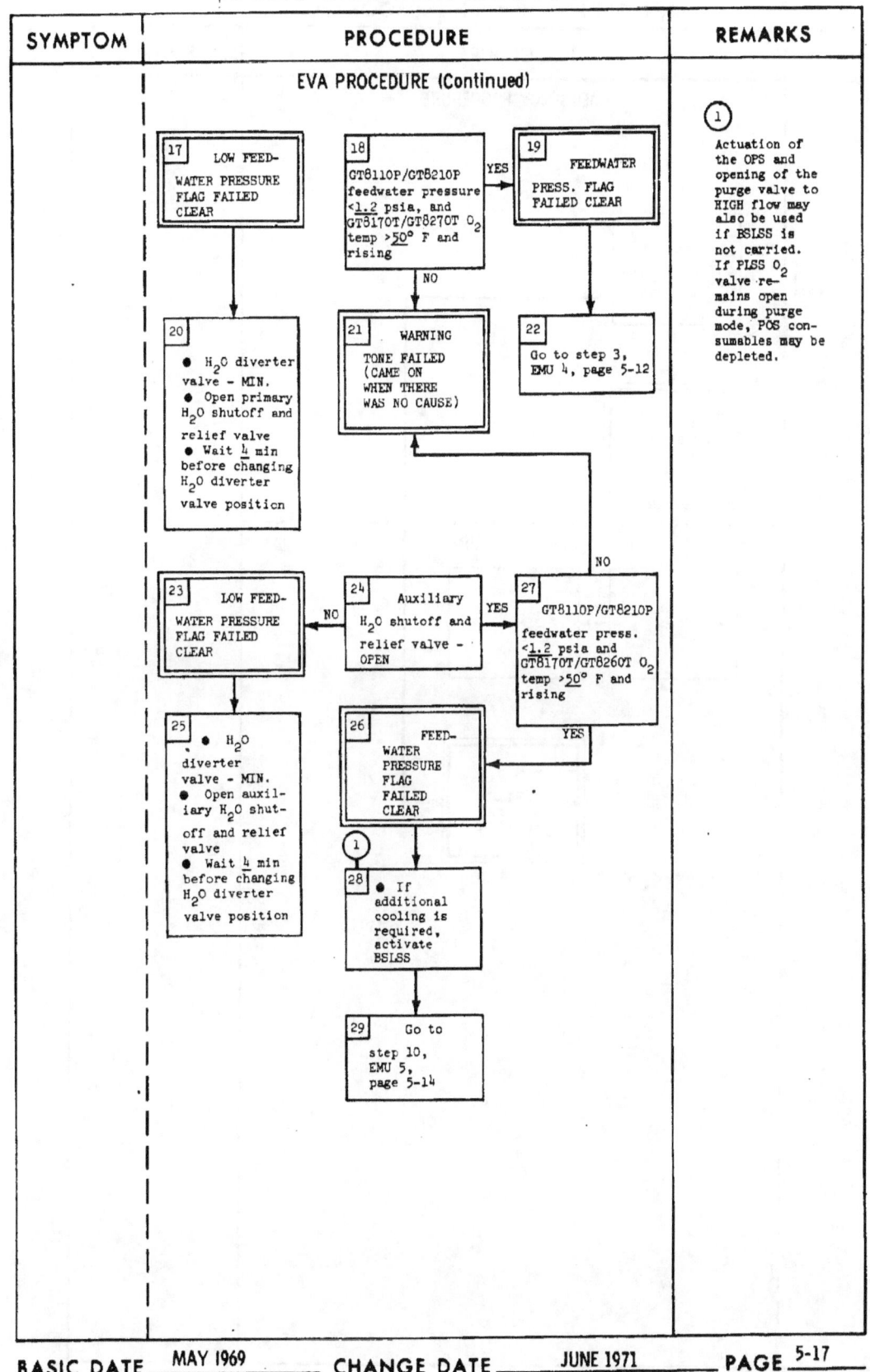

CSD-A-789-(2) III
APOLLO OPERATIONS HANDBOOK-EMU

SYMPTOM	PROCEDURE	REMARKS
EMU - 7 ① PGA pressure gage <3.7 psid and apparently stable and everything else normal Possible causes: a. Gage shift or failure b. PLSS O_2 regulator shift/ degraded	EMERGENCY PROCEDURE 1. • Actuate the OPS Does PGA pressure gage respond NO → 2. PGA PRESSURE GAGE FAILED YES → EVA PROCEDURE 3. Verify GT8168P/ GT8268P PGA pressure >3.7 psid → 4. • Turn off OPS 5. PLSS O_2 REGULATOR SHIFTED SET POINT	① PLSS O_2 pressure regulator regulates the PGA to 3.7 psid minimum if flow is 0.07 to 0.7 lb/hr.

BASIC DATE May 1969   CHANGE DATE JUNE 1971   PAGE 5-18

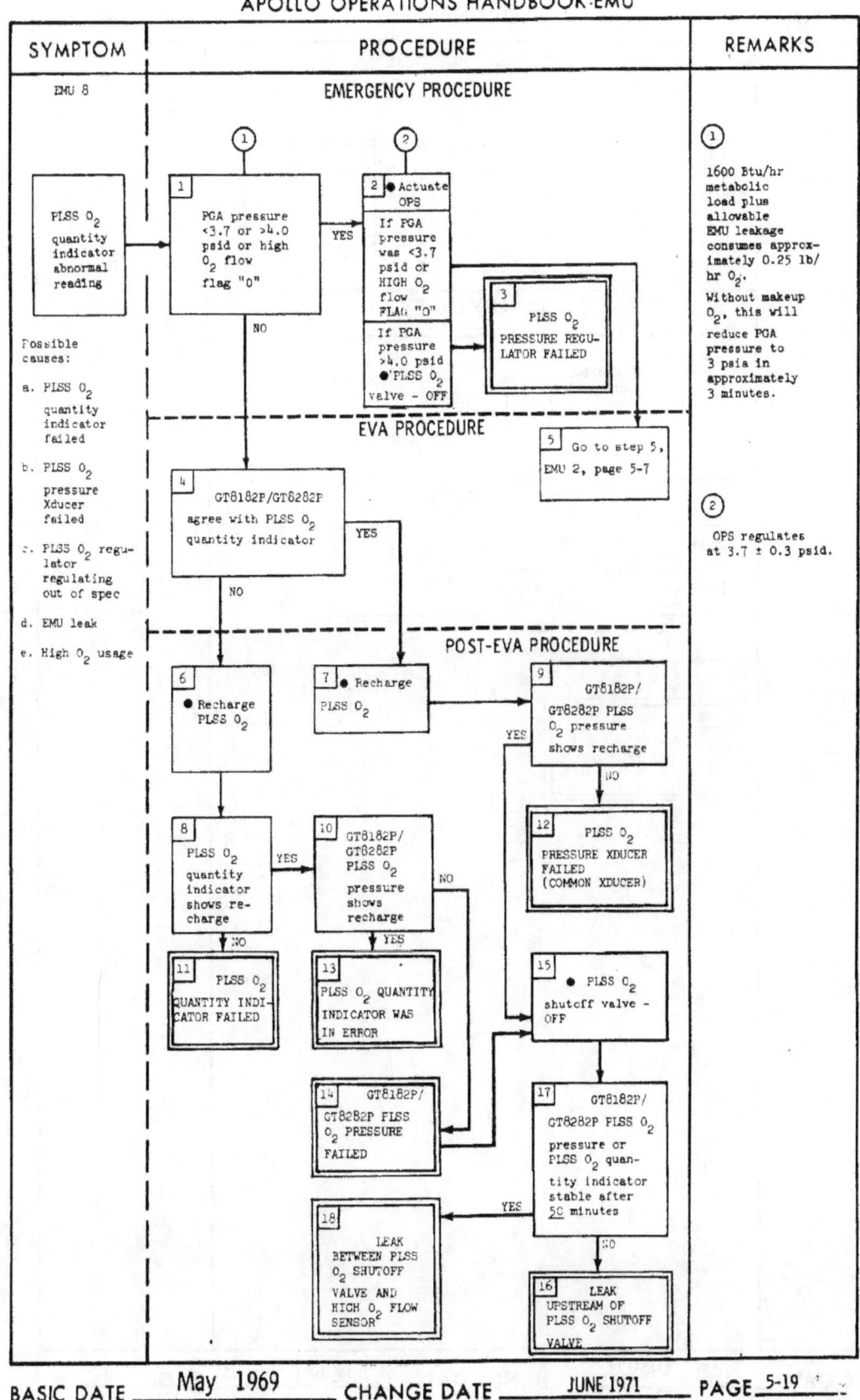

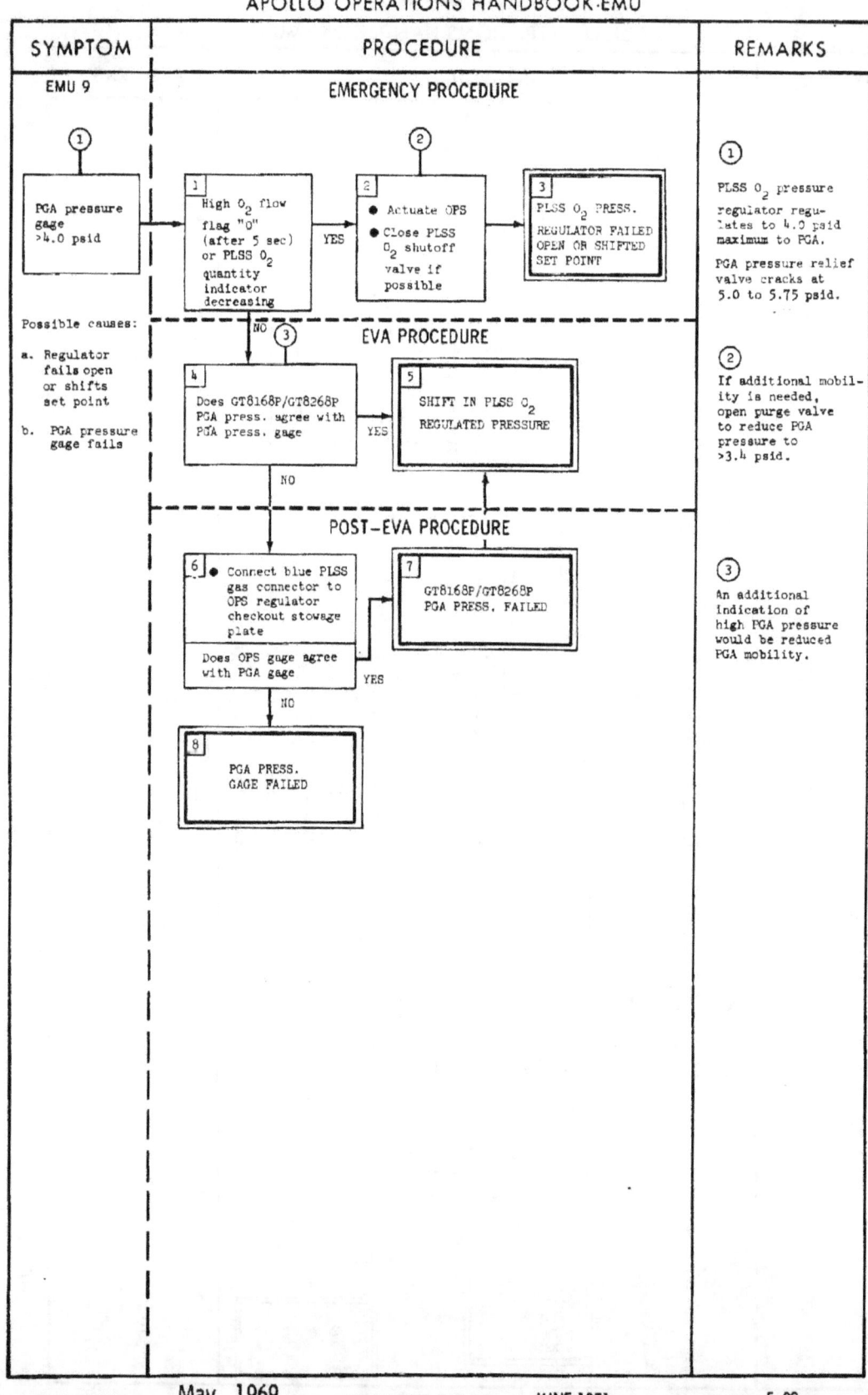

CSD-A-789-(2) III
APOLLO OPERATIONS HANDBOOK-EMU

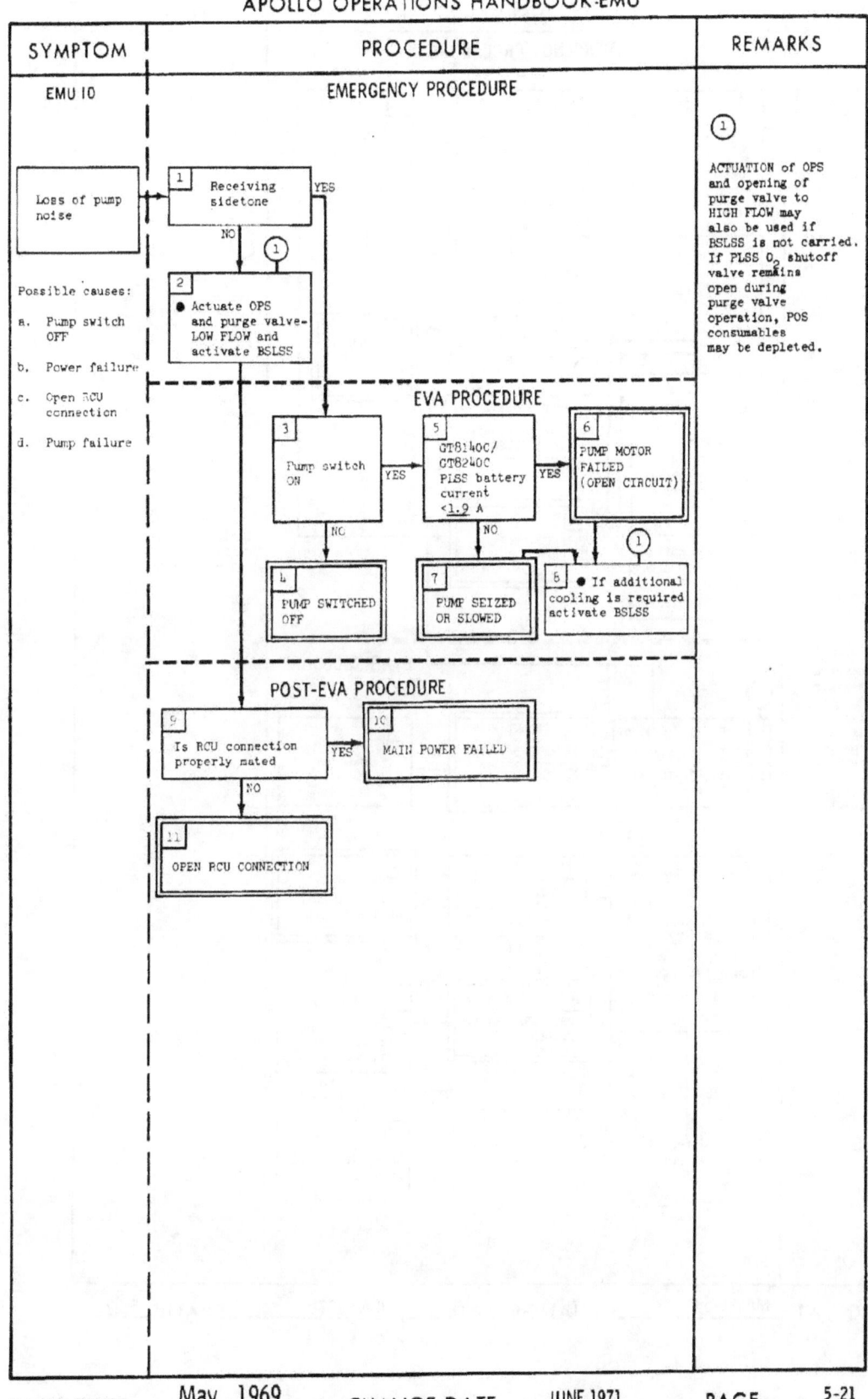

BASIC DATE May 1969   CHANGE DATE JUNE 1971   PAGE 5-21

CSD-A-789-(2) III
APOLLO OPERATIONS HANDBOOK-EMU

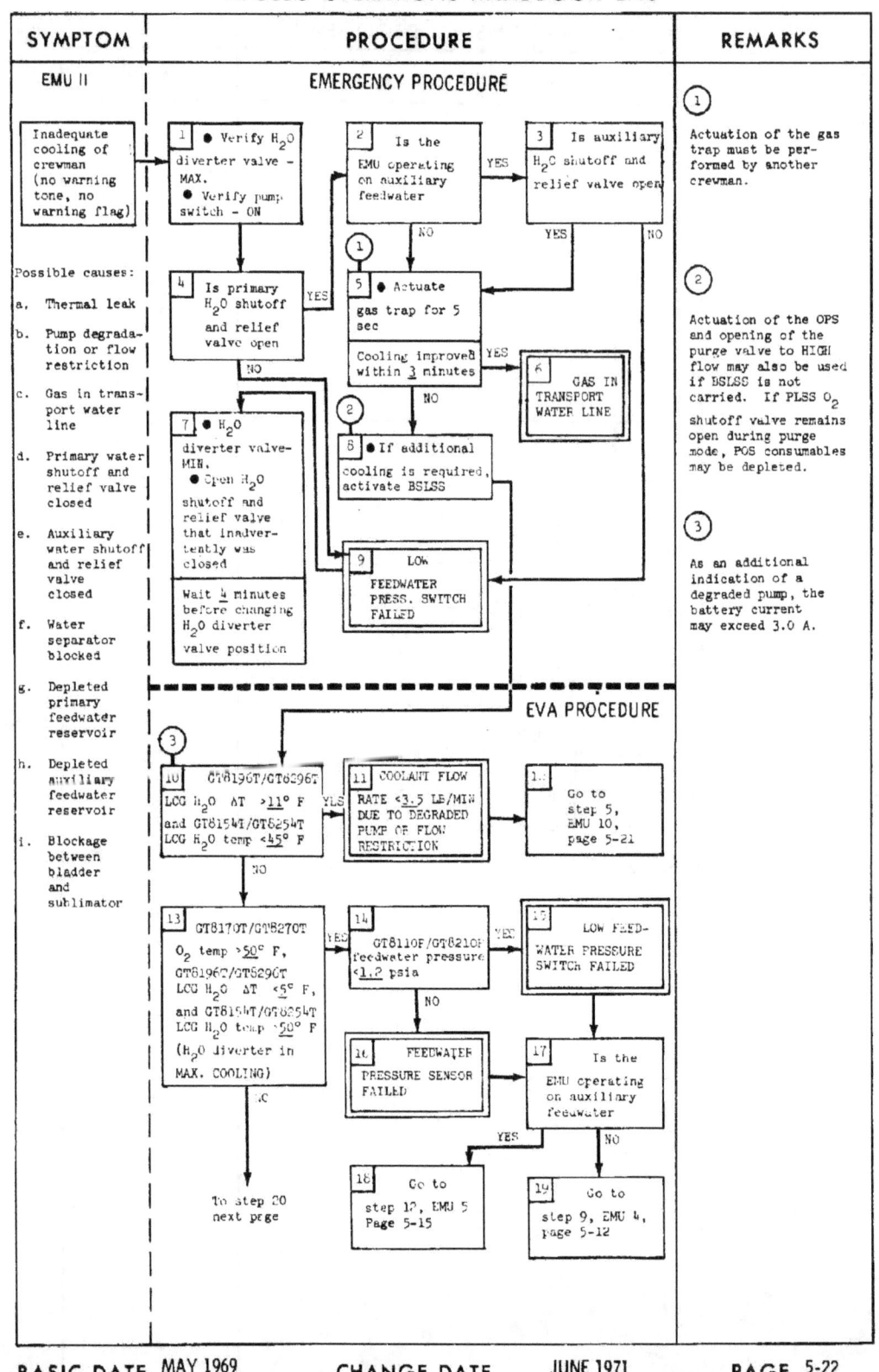

BASIC DATE MAY 1969   CHANGE DATE JUNE 1971   PAGE 5-22

SYMPTOM	PROCEDURE	REMARKS
	EVA PROCEDURE (Continued) 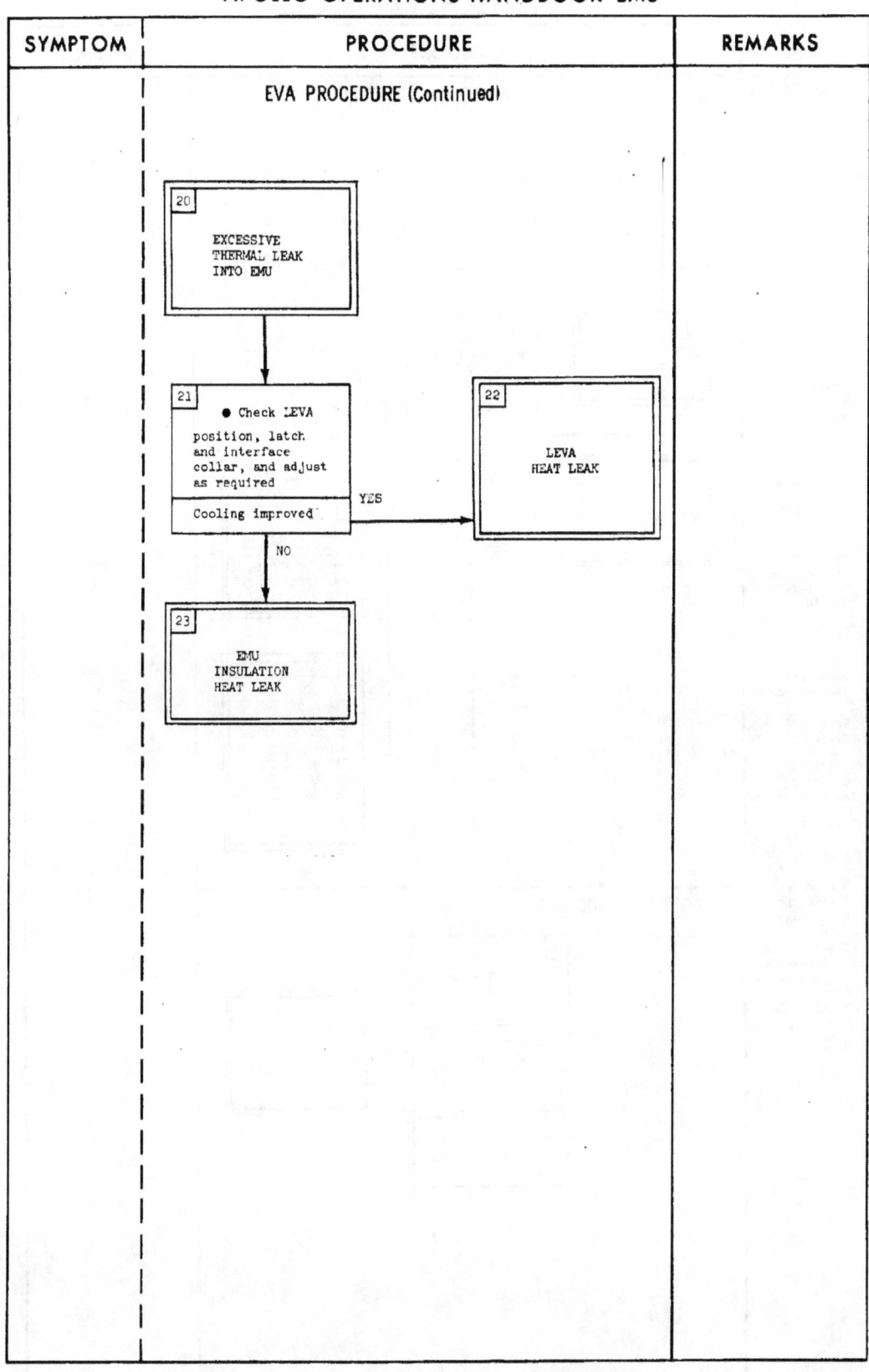	

CSD-A-789-(2) III
APOLLO OPERATIONS HANDBOOK-EMU

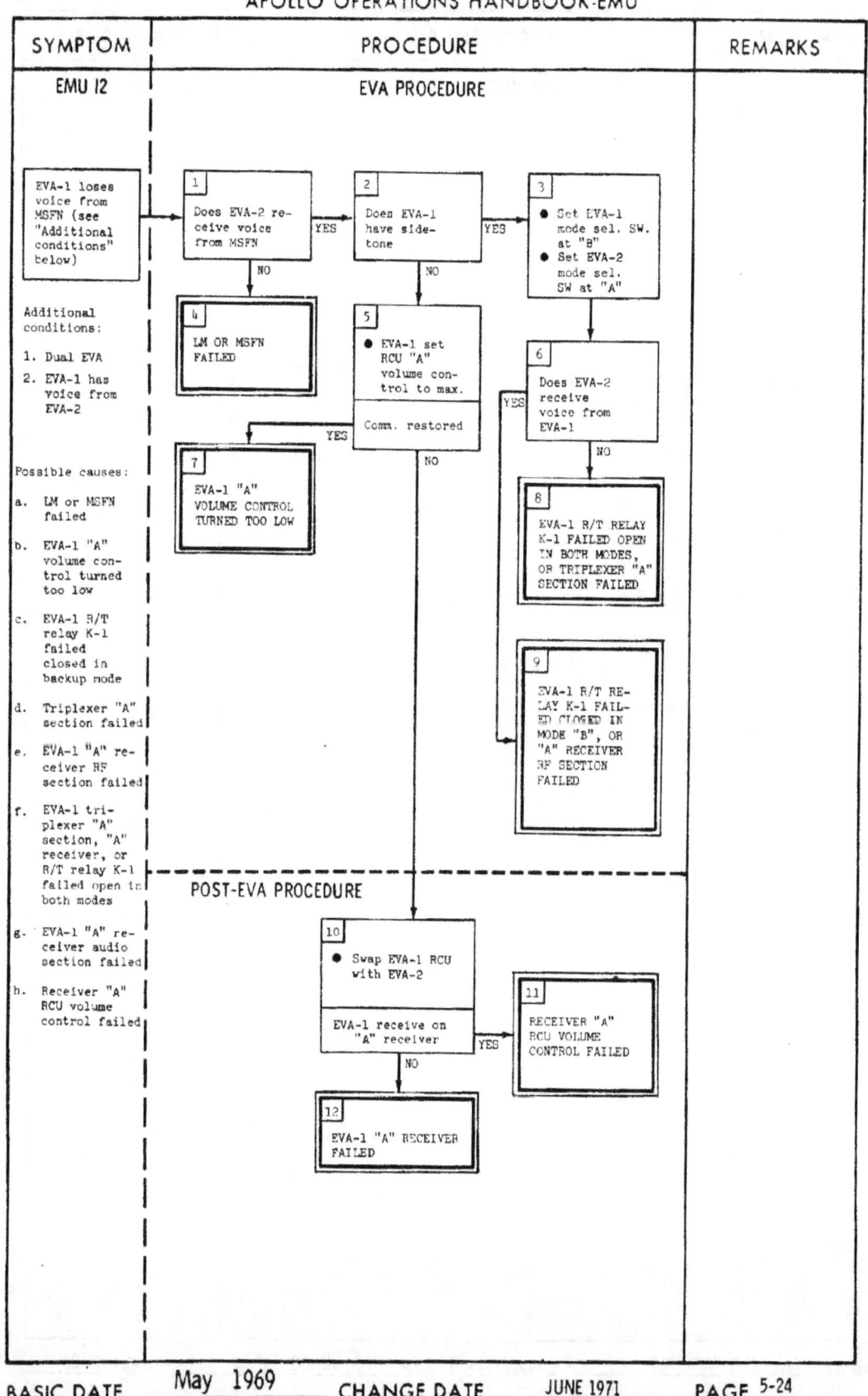

BASIC DATE May 1969   CHANGE DATE JUNE 1971   PAGE 5-24

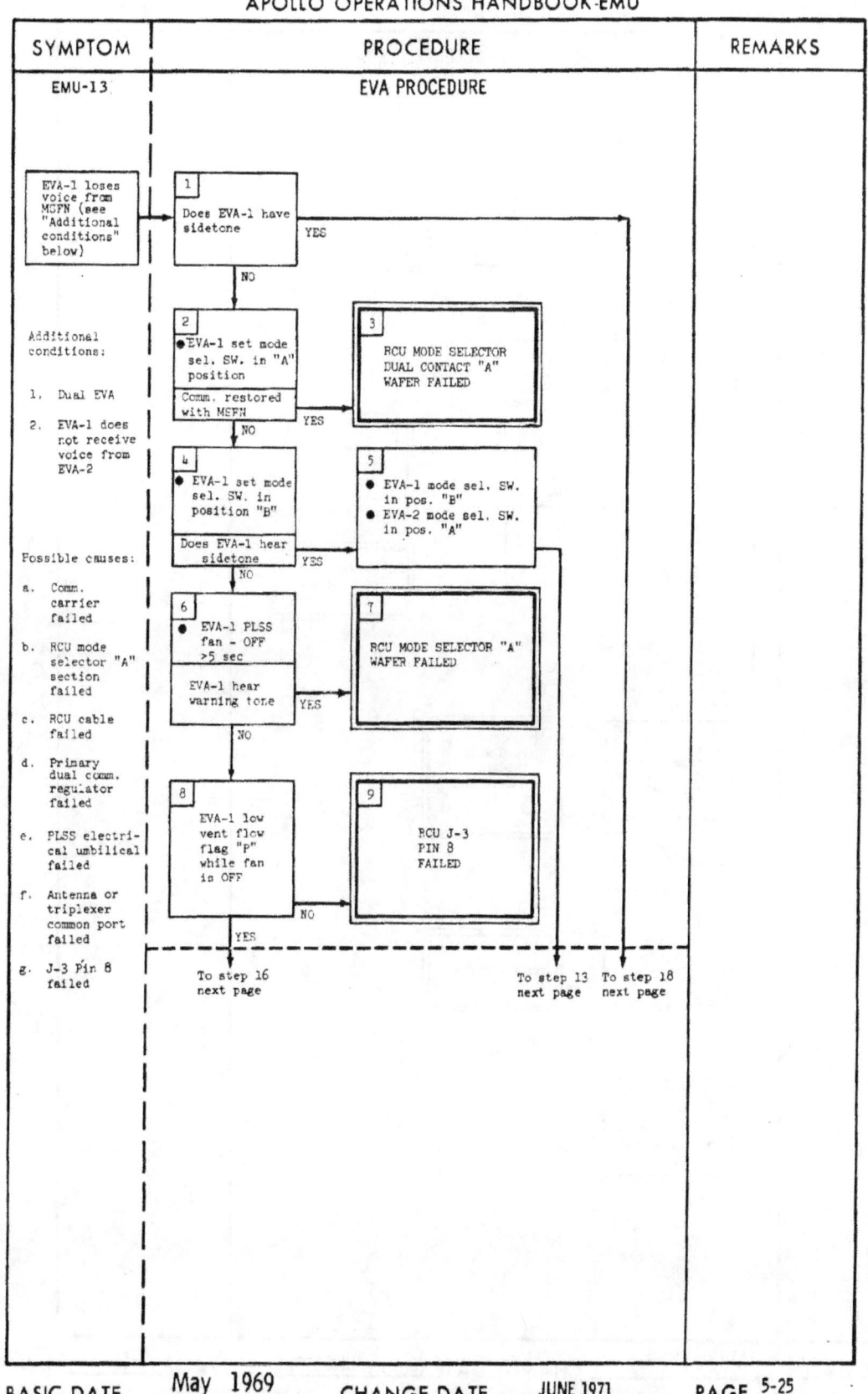

CSD-A-789-(2) III
APOLLO OPERATIONS HANDBOOK-EMU

SYMPTOM	PROCEDURE	REMARKS
	POST-EVA PROCEDURE 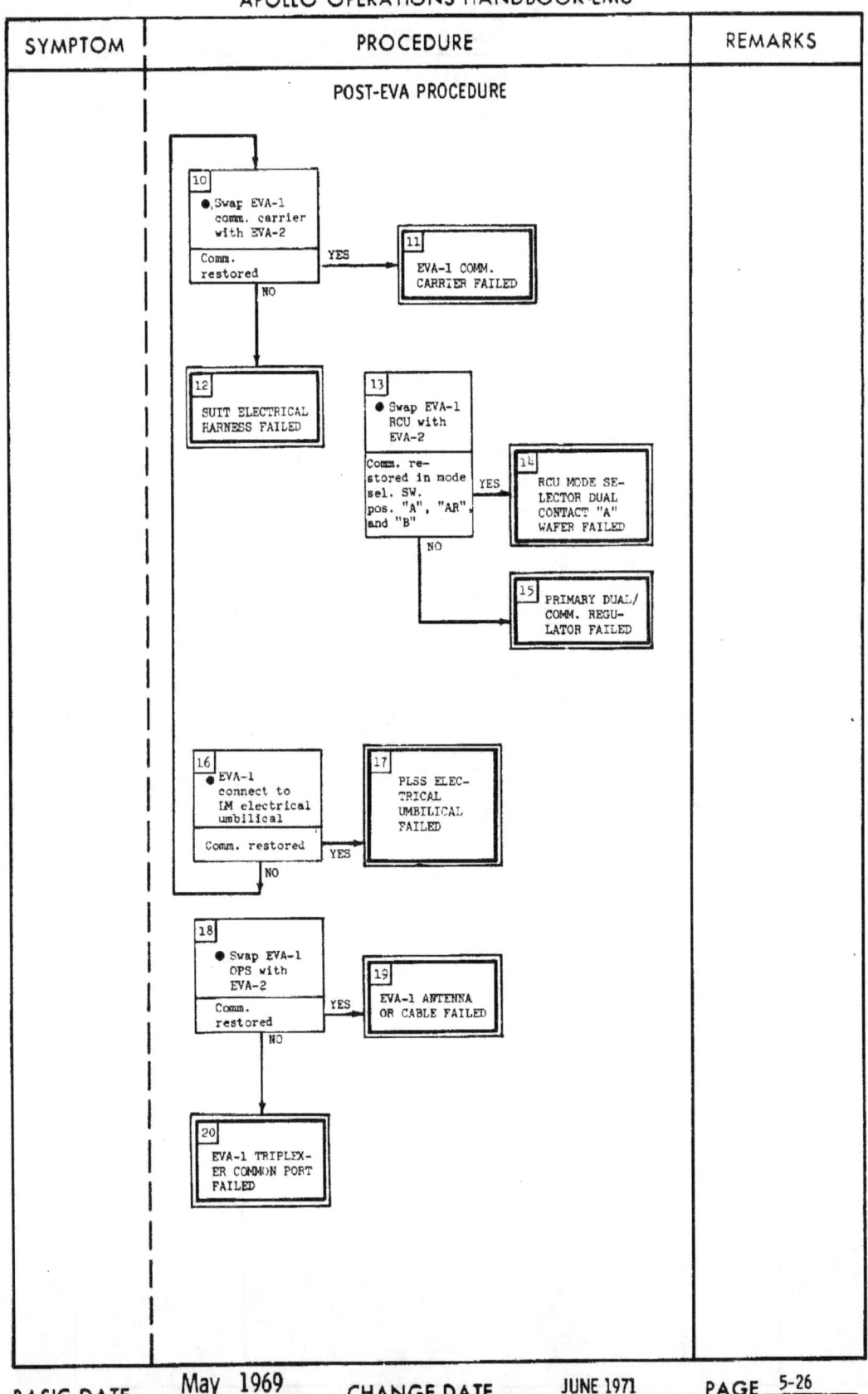	

BASIC DATE May 1969 CHANGE DATE JUNE 1971 PAGE 5-26

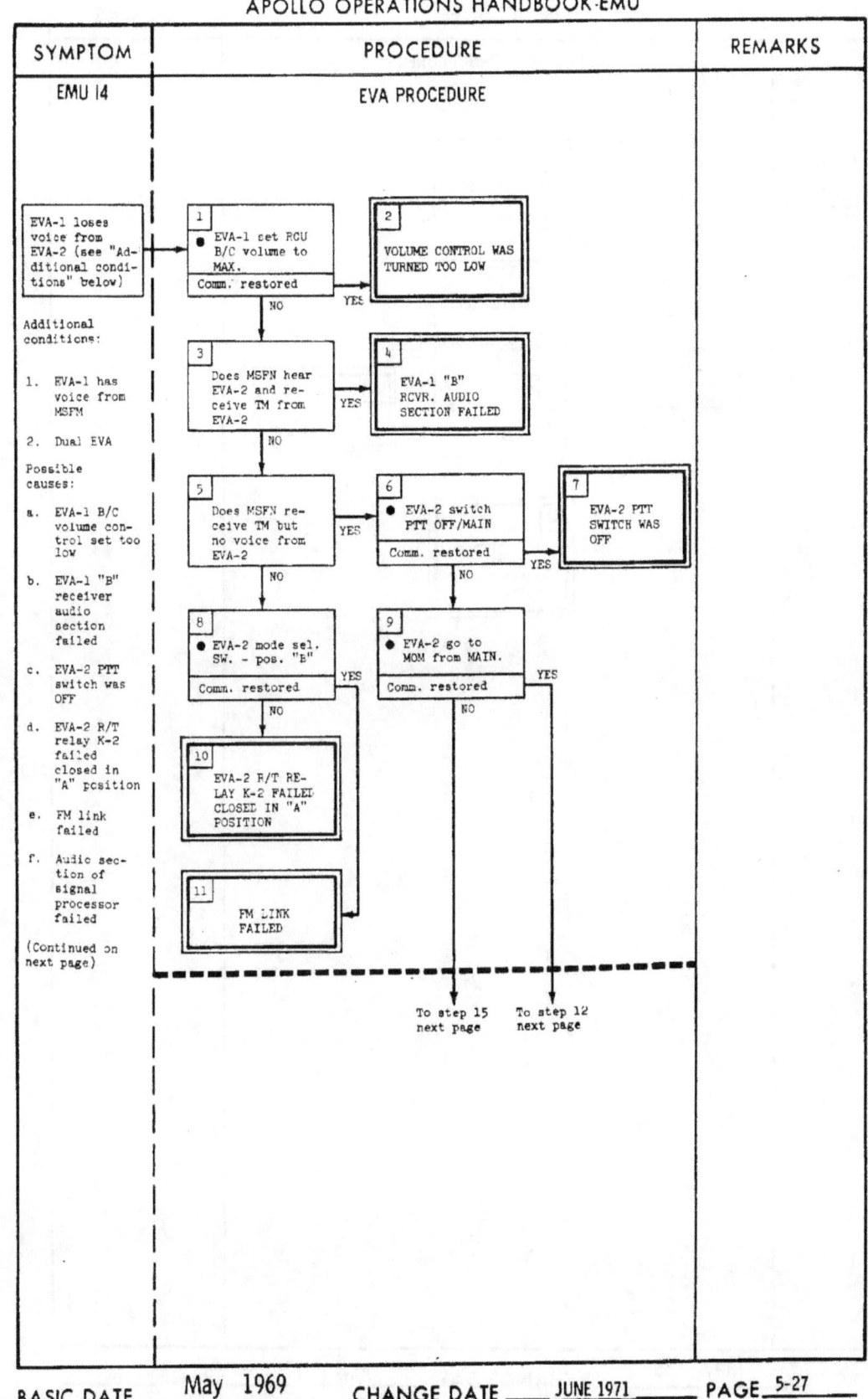

SYMPTOM	PROCEDURE	REMARKS
	POST-EVA PROCEDURE	

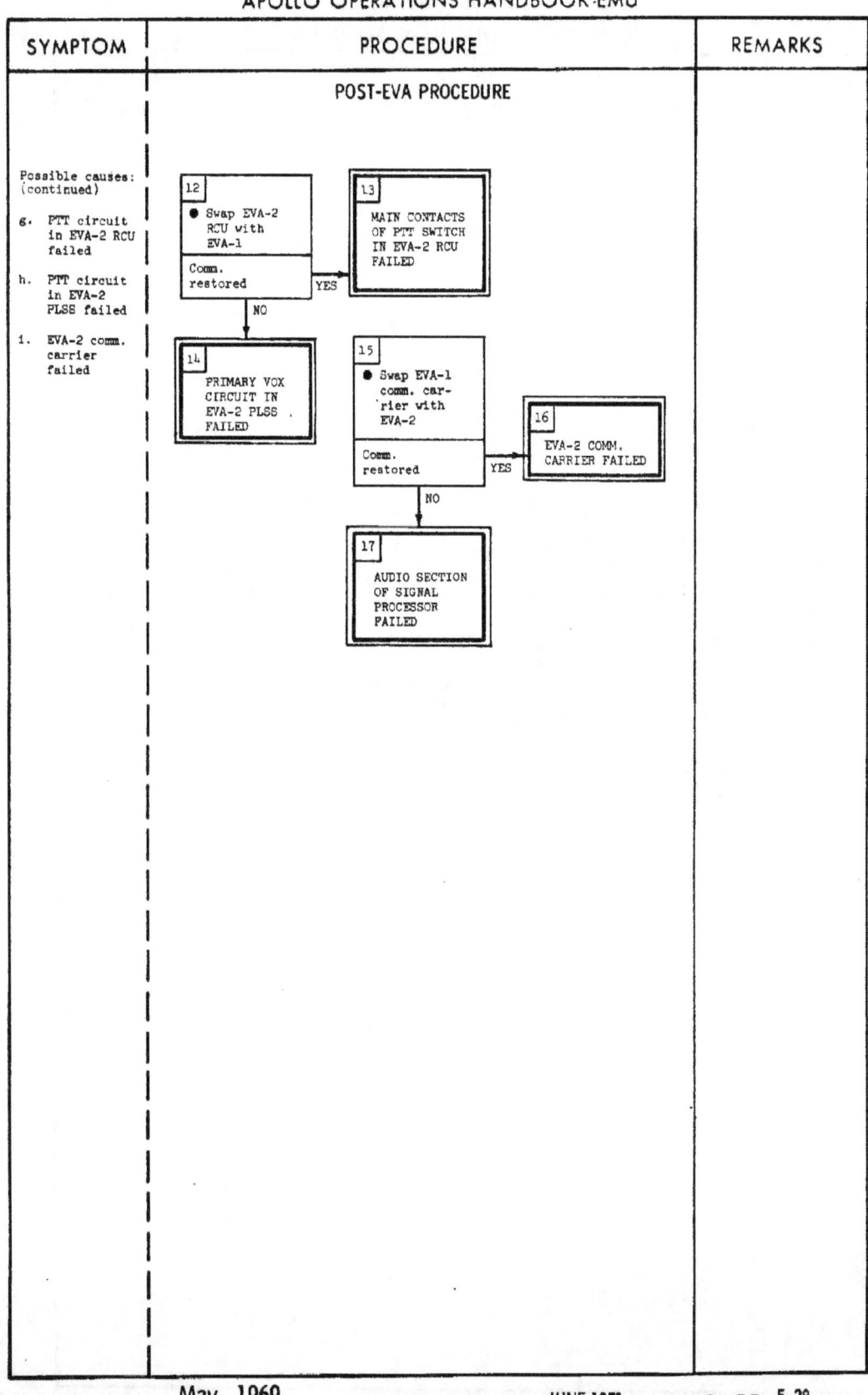

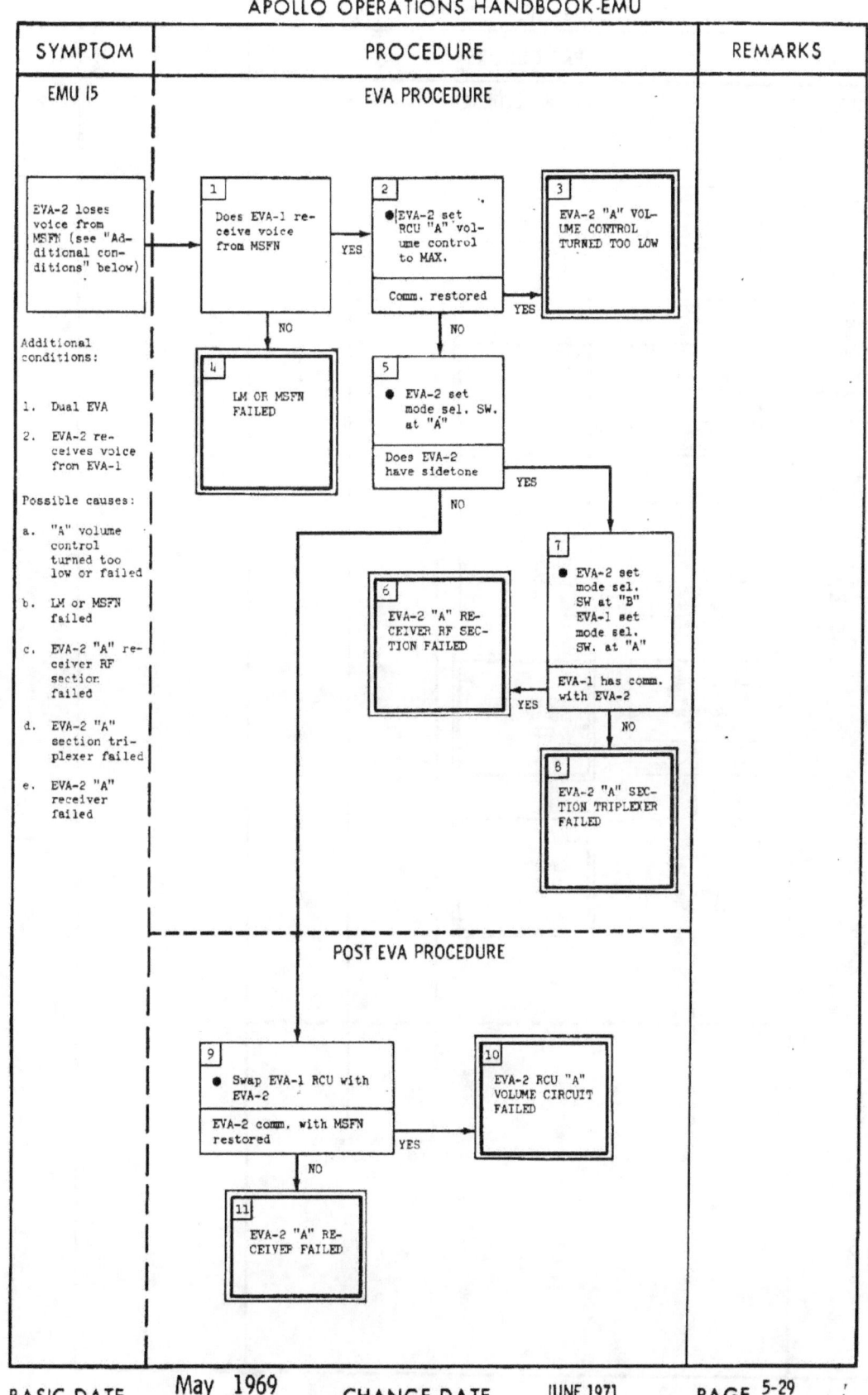

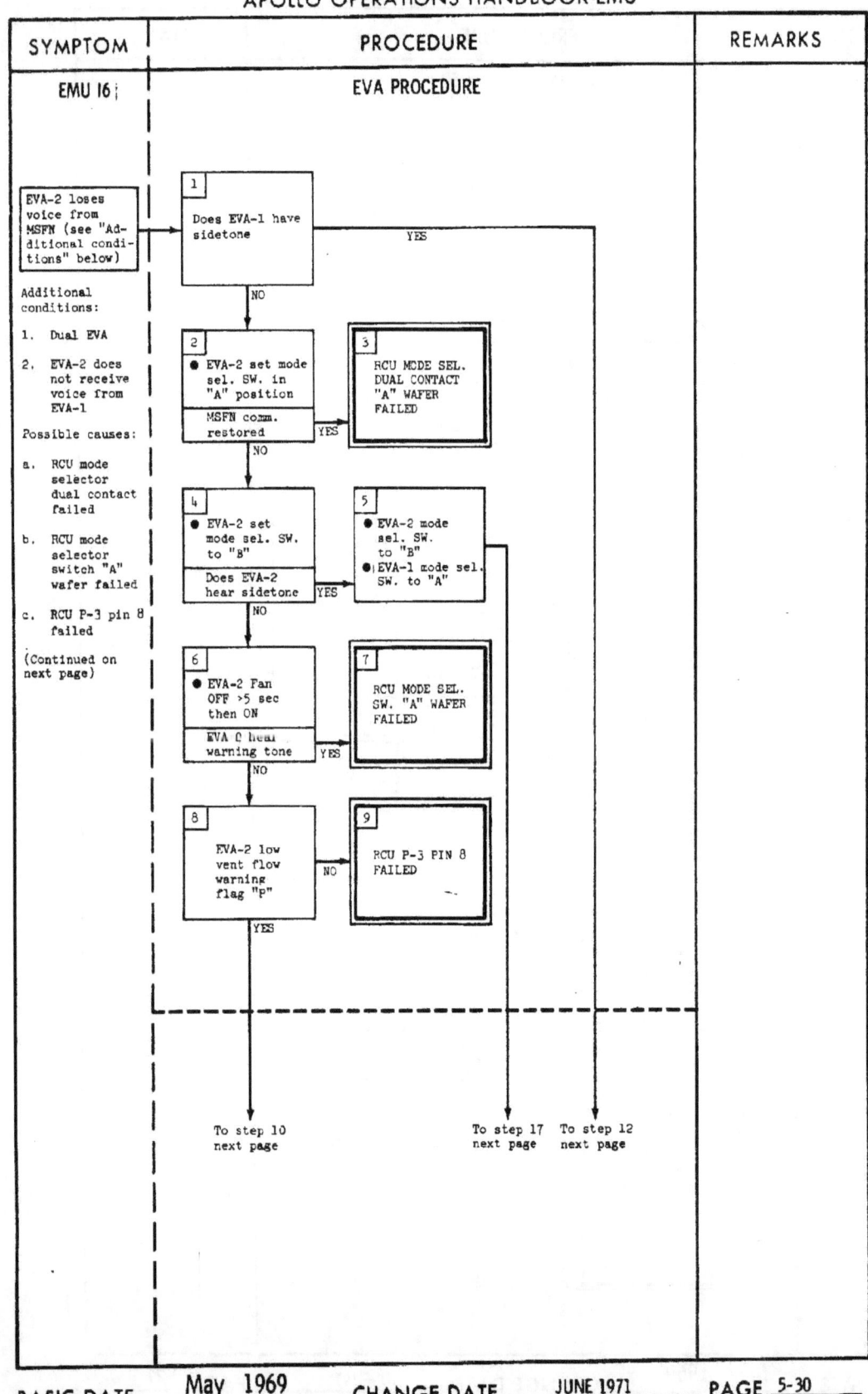

CSD-A-789-(2) III
APOLLO OPERATIONS HANDBOOK-EMU

SYMPTOM	PROCEDURE	REMARKS
d. PLSS electrical umbilical failed e. EVA-2 comm. carrier failed f. EVA-2 triplexer failed g. EVA-2 suit electrical harness failed h. RCU mode selector switch dual and primary contact "A" wafer failed i. Primary/dual comm. regulator failed j. EVA-2 antenna or cable failed	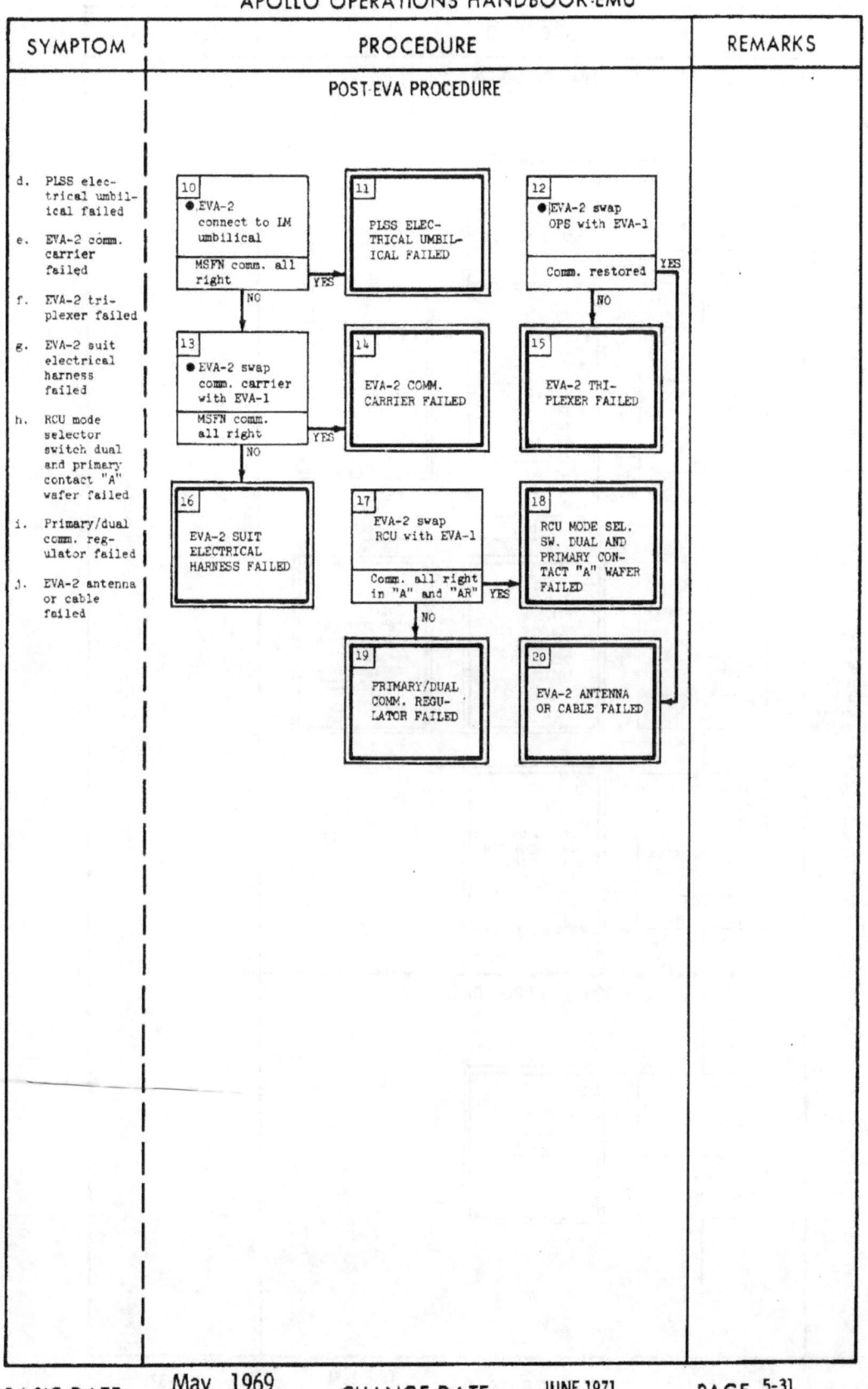	

BASIC DATE May 1969   CHANGE DATE JUNE 1971   PAGE 5-31

CSD-A-789-(2) III
APOLLO OPERATIONS HANDBOOK-EMU

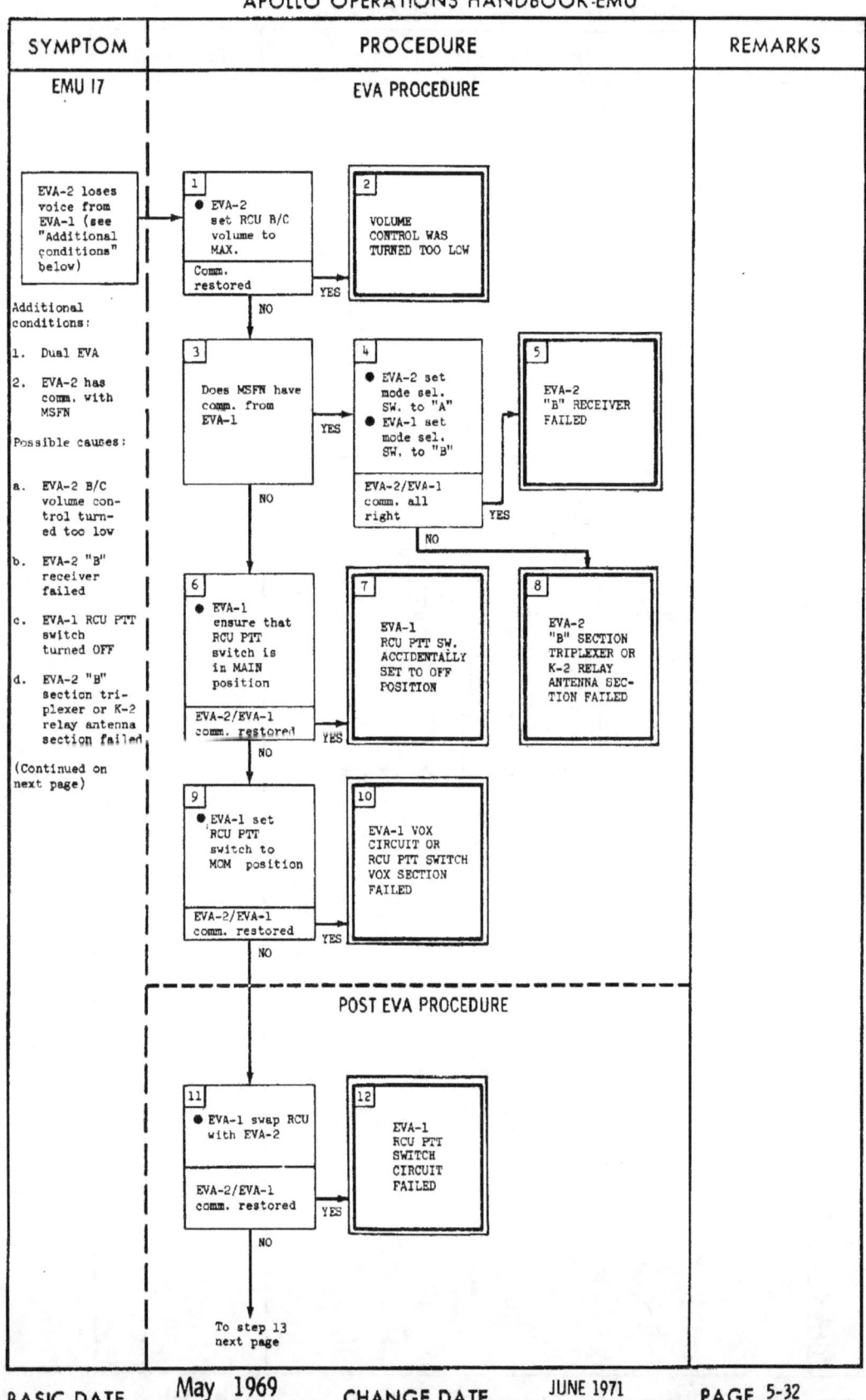

BASIC DATE: May 1969 — CHANGE DATE: JUNE 1971 — PAGE 5-32

APOLLO OPERATIONS HANDBOOK-EMU

SYMPTOM	PROCEDURE	REMARKS
Possible causes: (continued) e. EVA-1 VOX circuit or RCU PTT switch VOX section failed f. EVA-1 RCU PTT switch circuit failed g. EVA-1 primary/dual signal processor failed h. EVA-1 VOX/PTT circuit failed i. EVA-1 comm. carrier failed j. EVA-1 suit electrical harness failed	POST-EVA PROCEDURE (Continued) 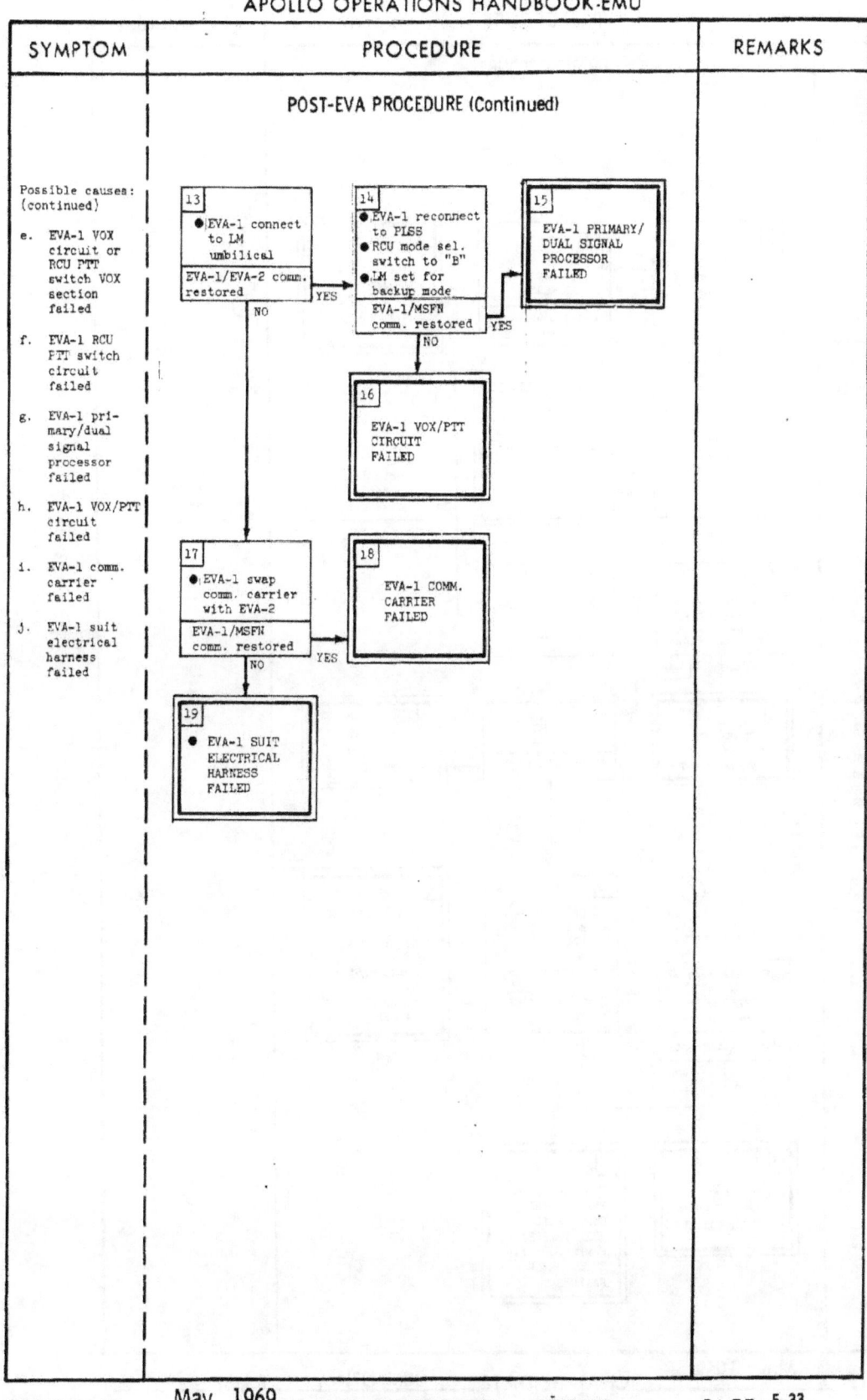	

PAGE 5-33

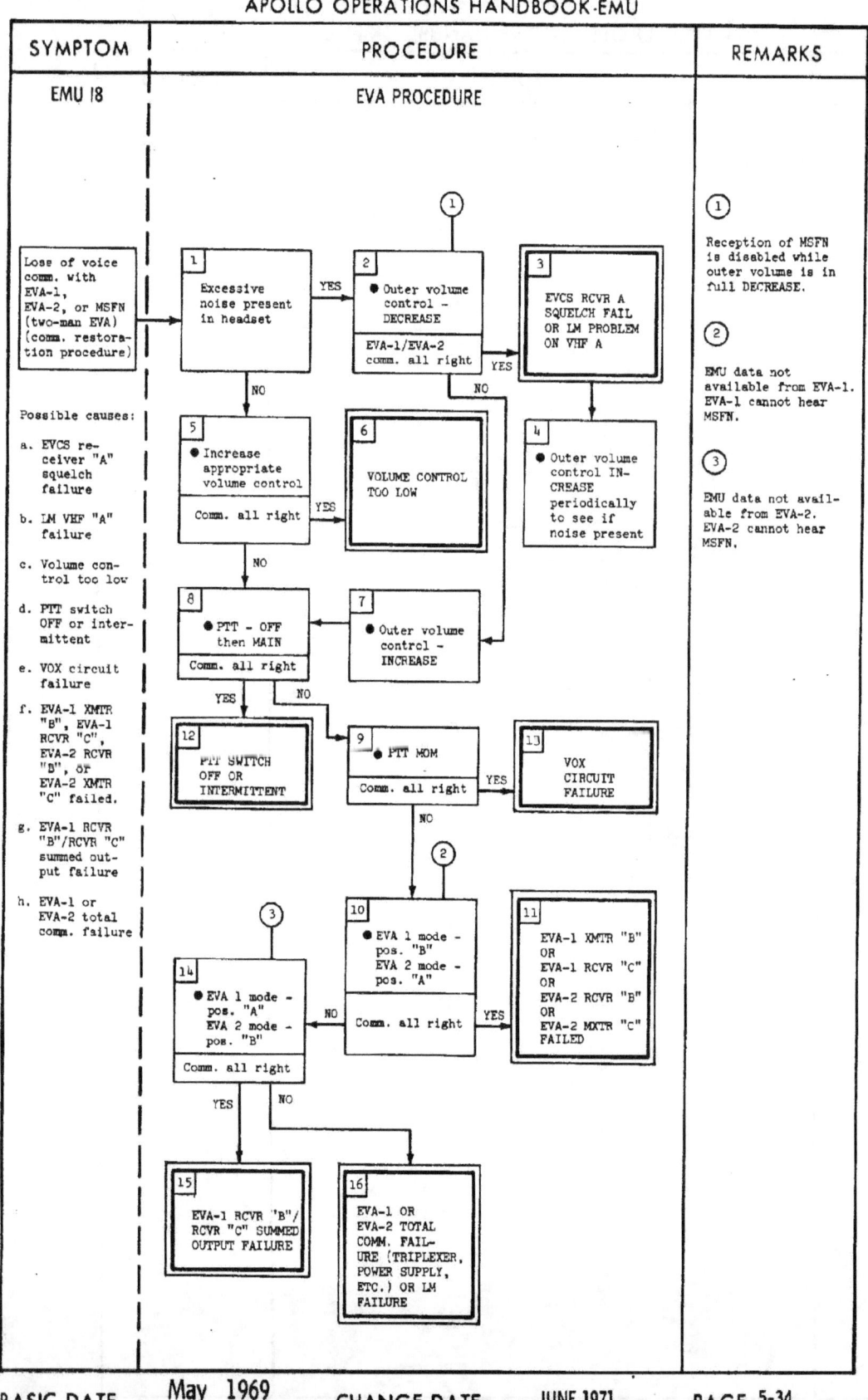

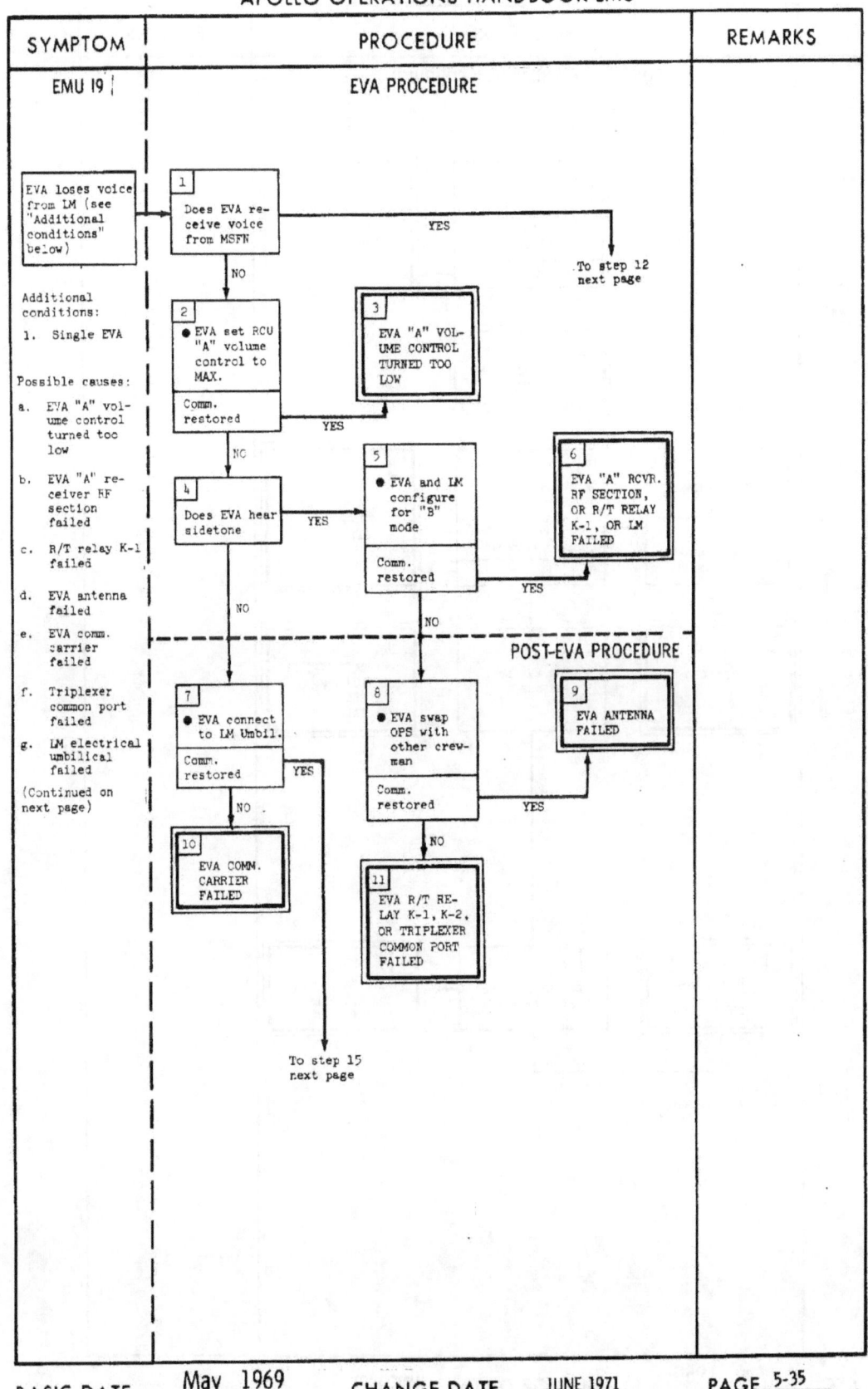

CSD-A-789-(2) III
APOLLO OPERATIONS HANDBOOK-EMU

SYMPTOM	PROCEDURE	REMARKS

POST-EVA PROCEDURE (Continued)

Possible causes: (continued)

h. EVA RCU "A" volume circuit failed

i. LM failed

j. EVA "A" receiver RF section failed

k. PLSS electrical umbilical failed

l. Primary dual comm. regulator failed

m. EVA "A" receiver failed

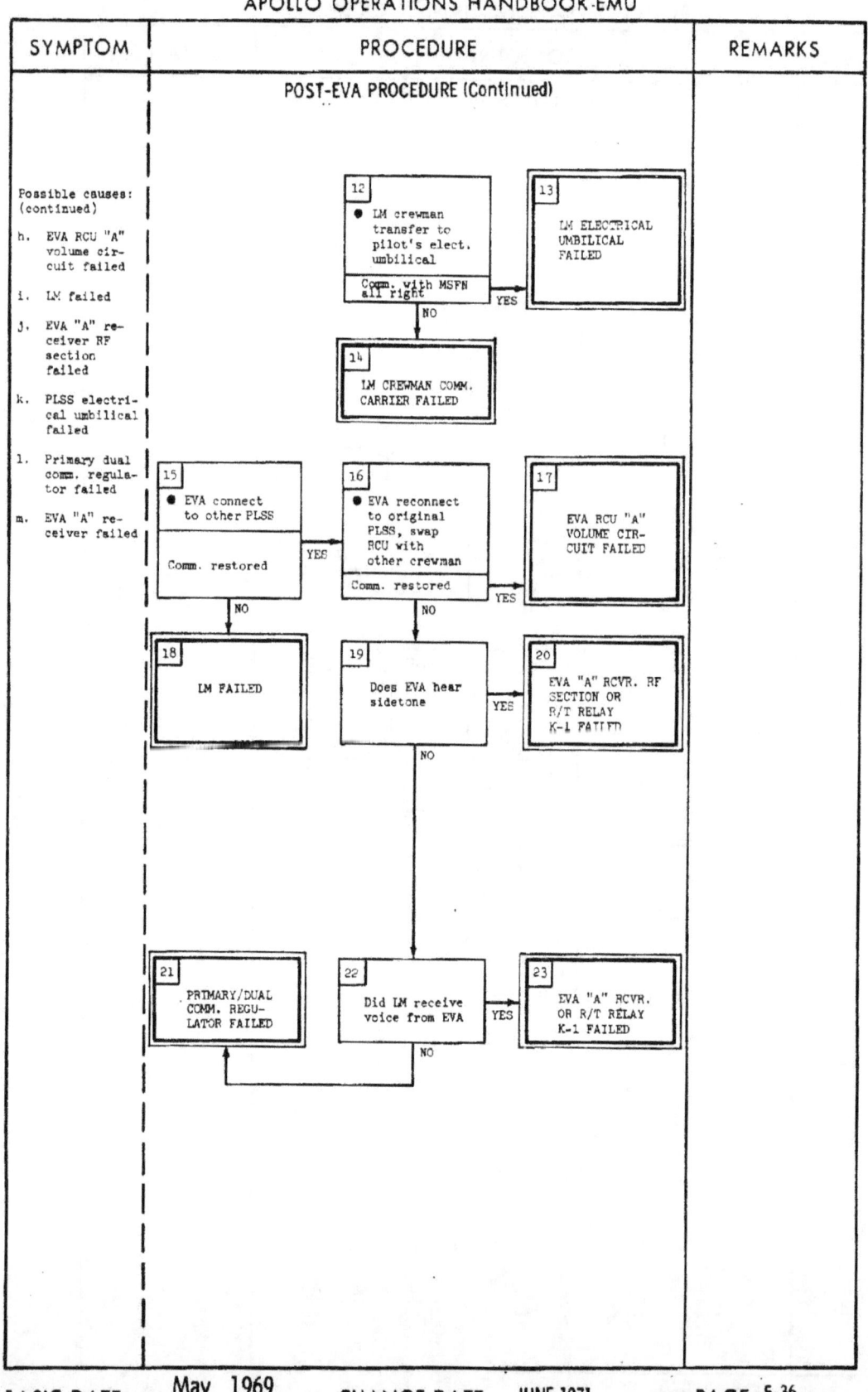

BASIC DATE May 1969   CHANGE DATE JUNE 1971   PAGE 5-36

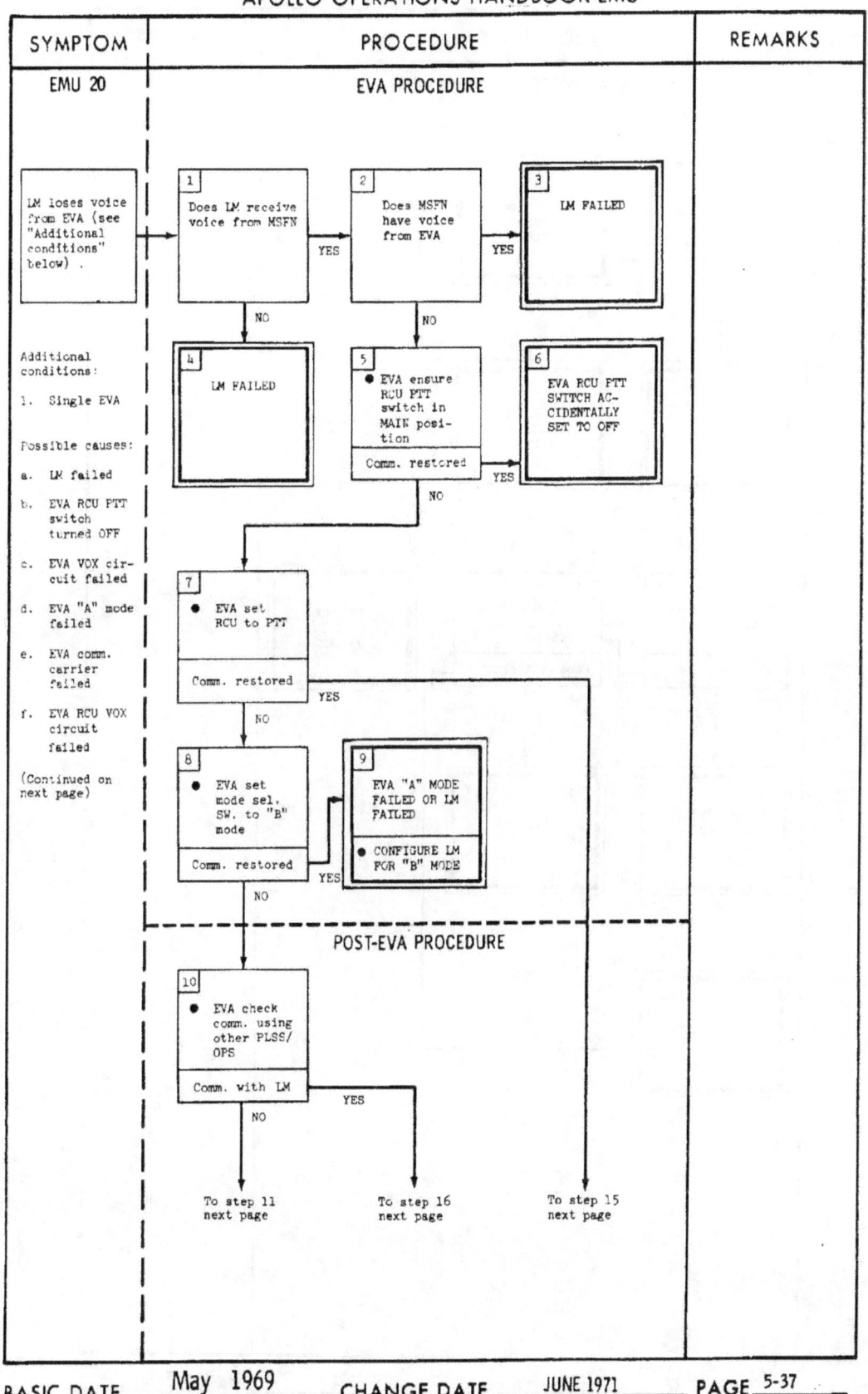

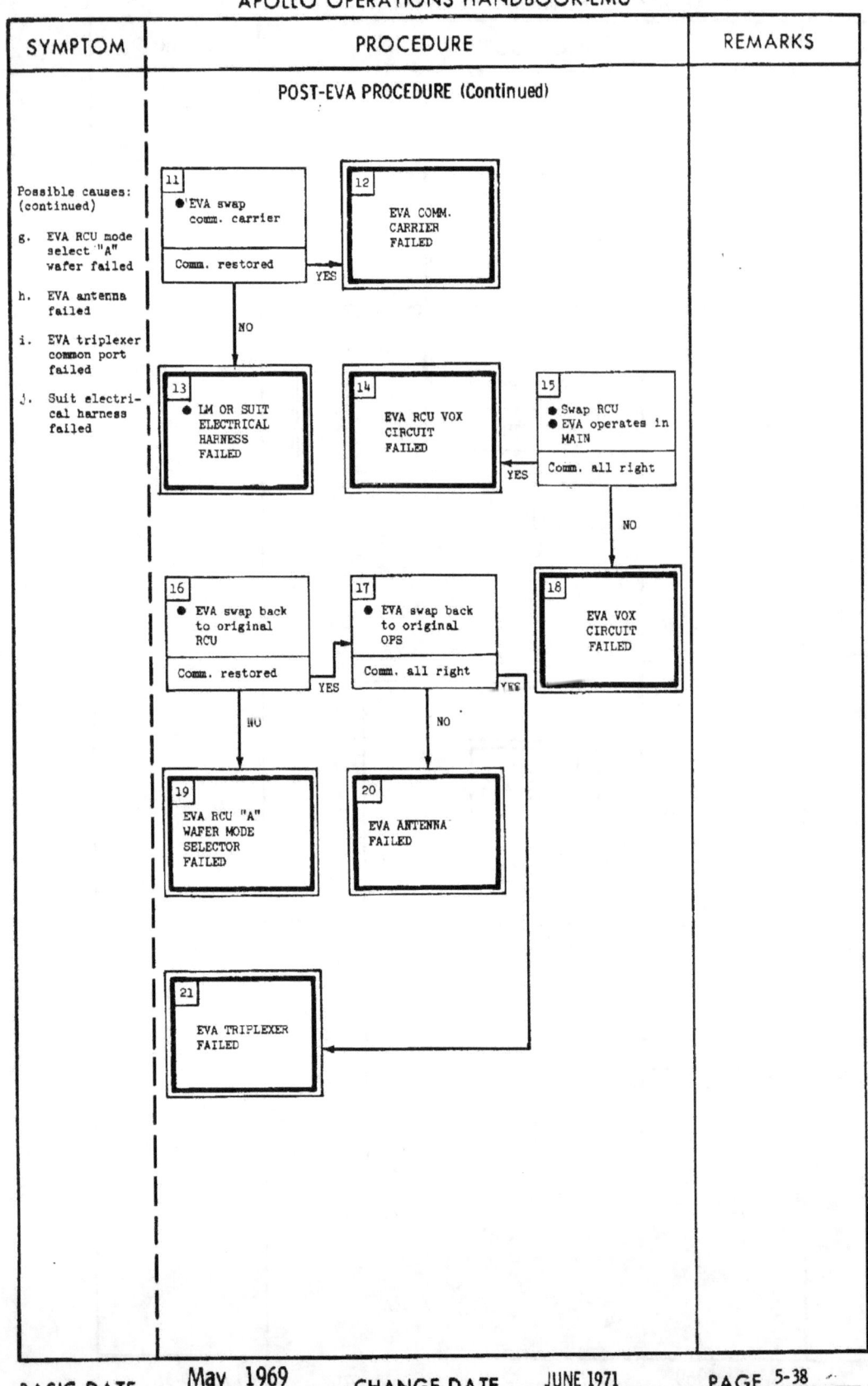

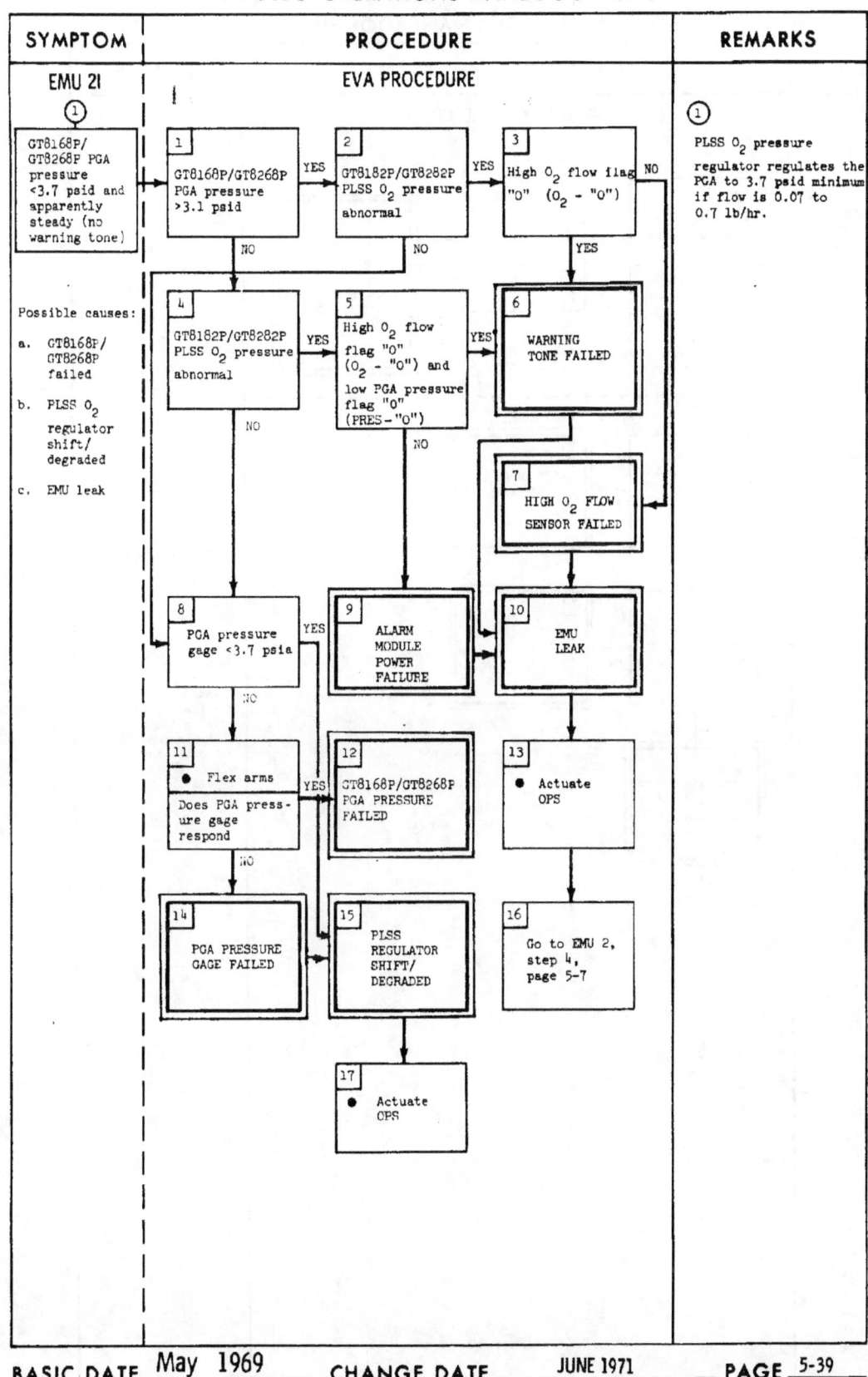

CSD-A-789-(2) III
APOLLO OPERATIONS HANDBOOK-EMU

SYMPTOM	PROCEDURE	REMARKS
EMU 22		

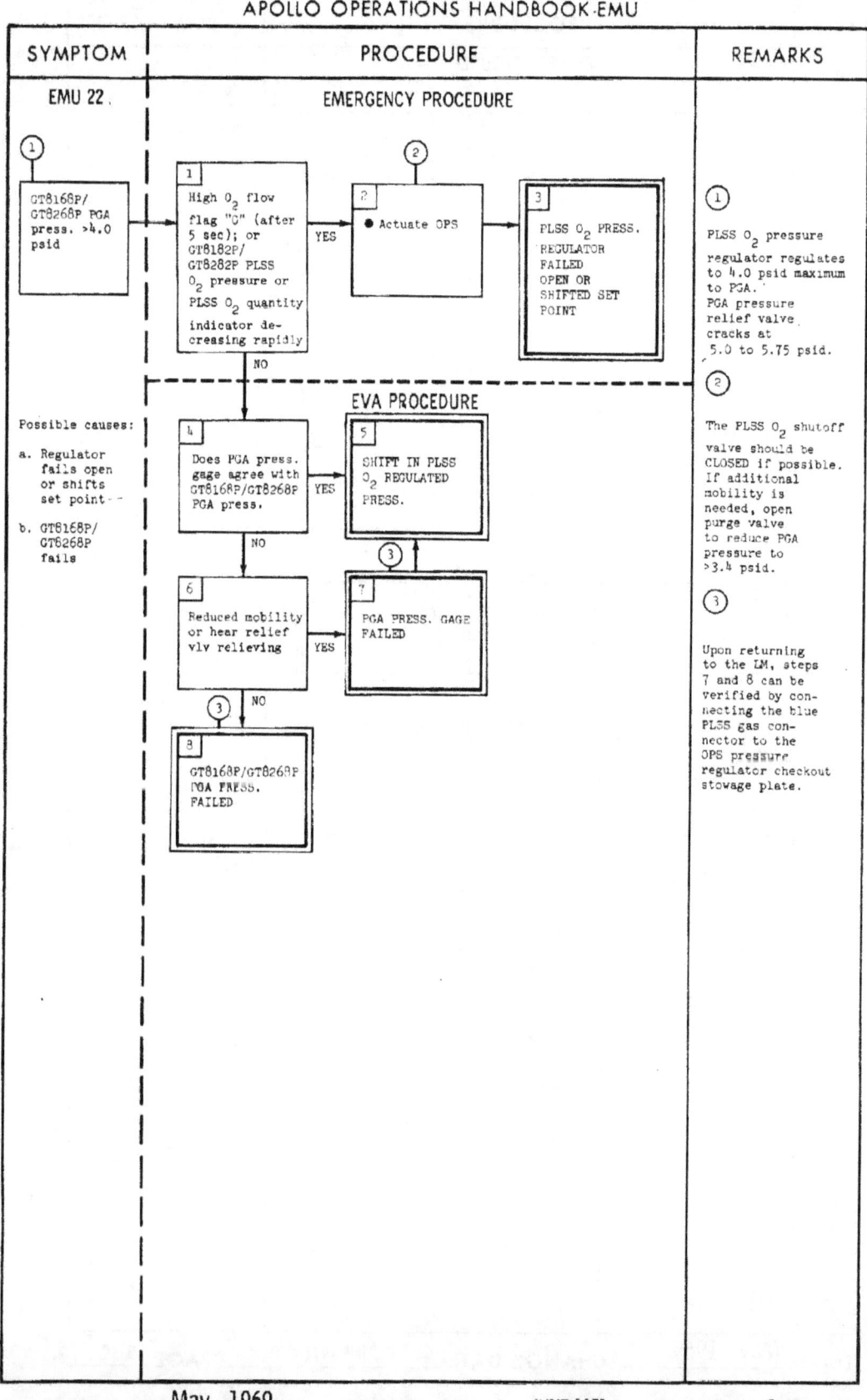

BASIC DATE: May 1969   CHANGE DATE: JUNE 1971   PAGE 5-40

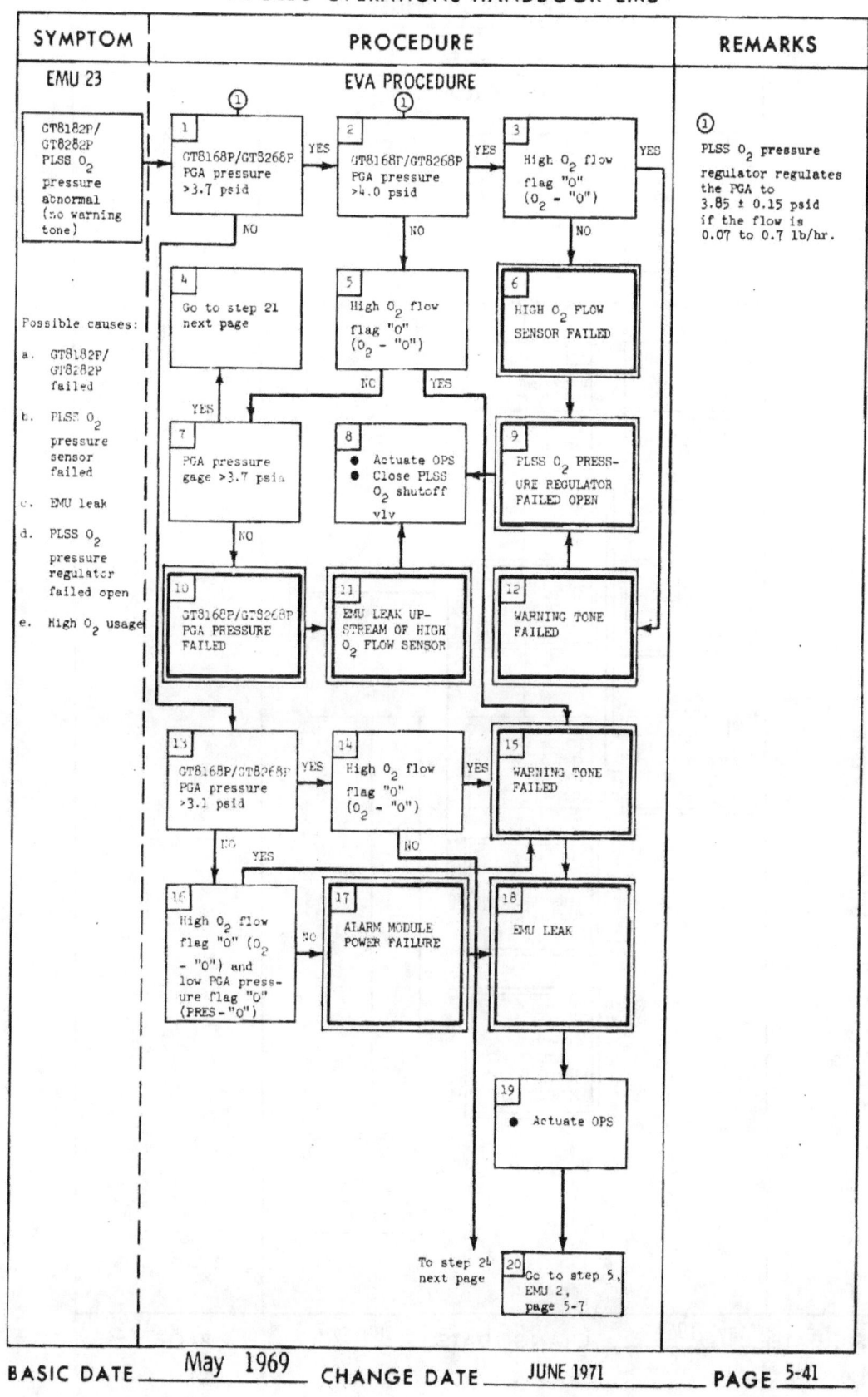

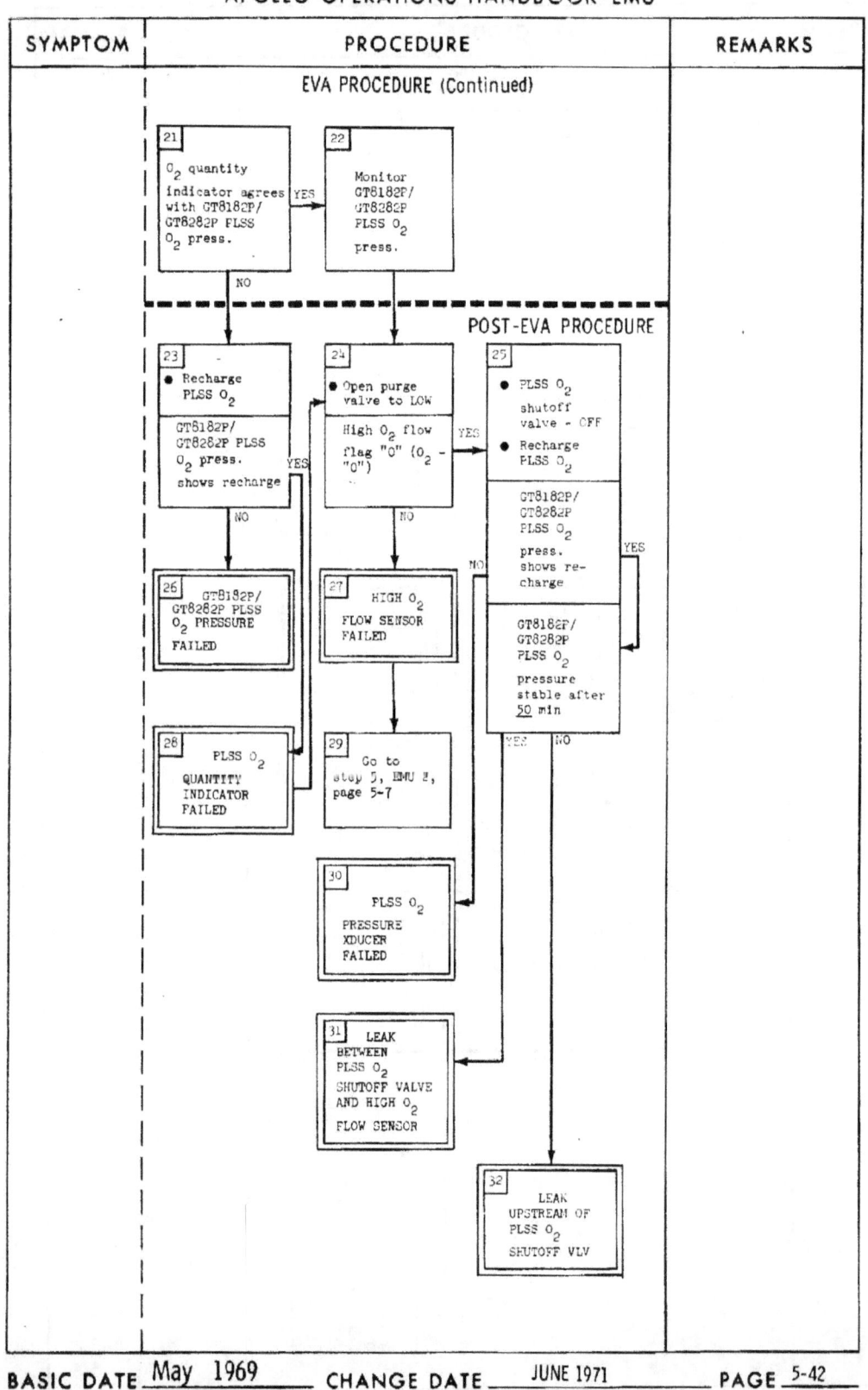

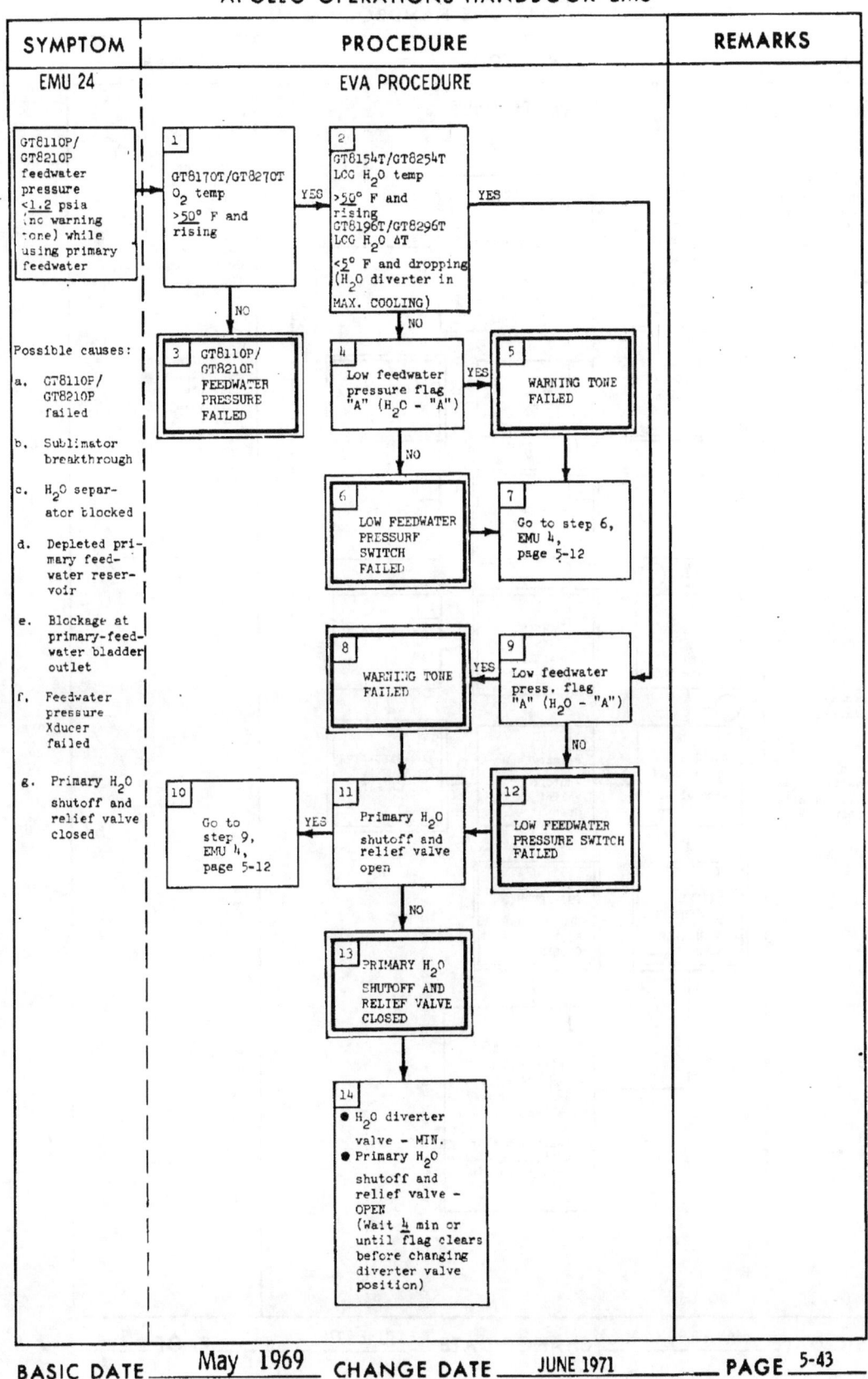

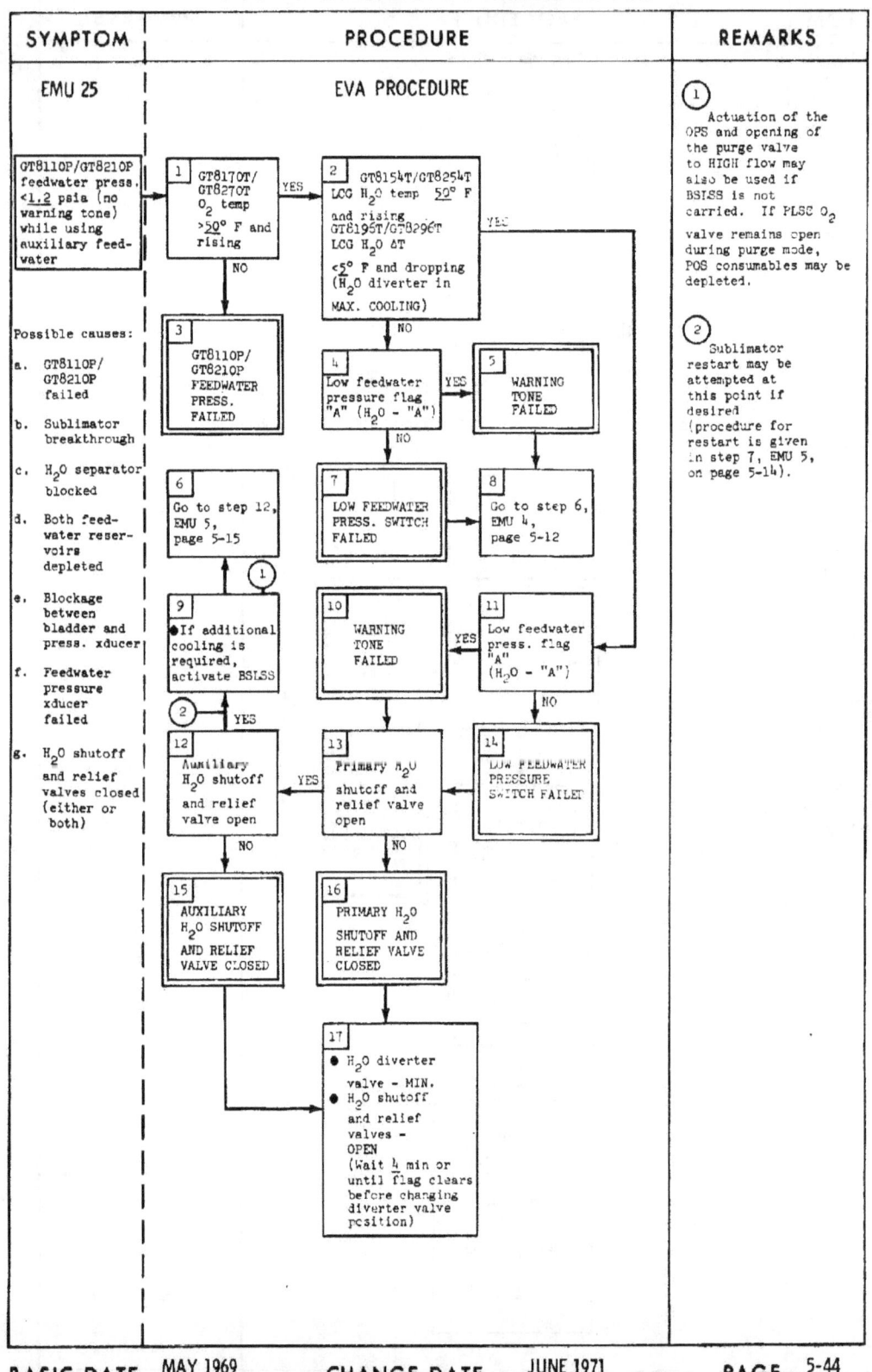

CSD-A-789-(2) III
APOLLO OPERATIONS HANDBOOK - EMU

SYMPTOM	PROCEDURE	REMARKS
	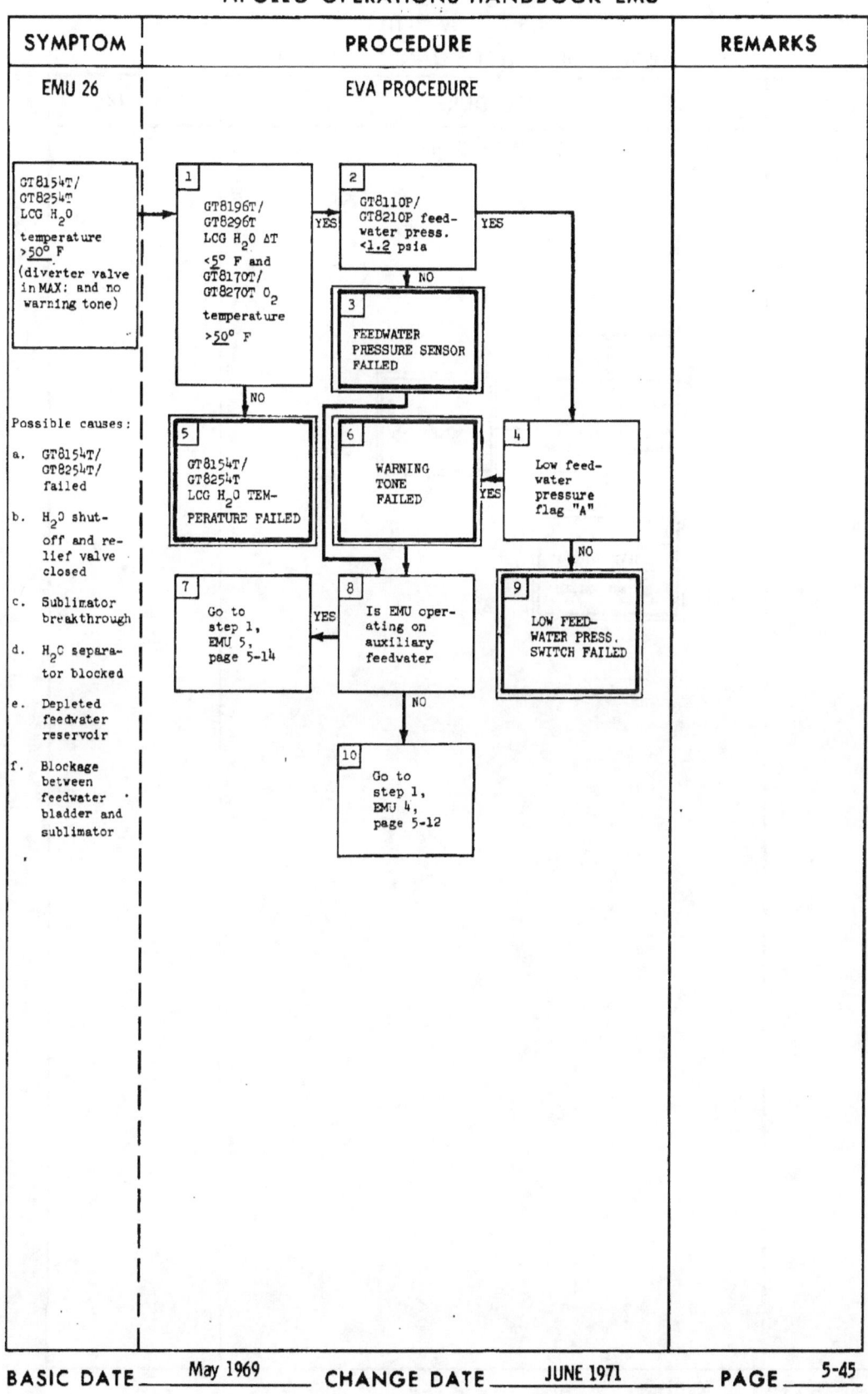	

PAGE 5-45

SYMPTOM	PROCEDURE	REMARKS
	POST-EVA PROCEDURE 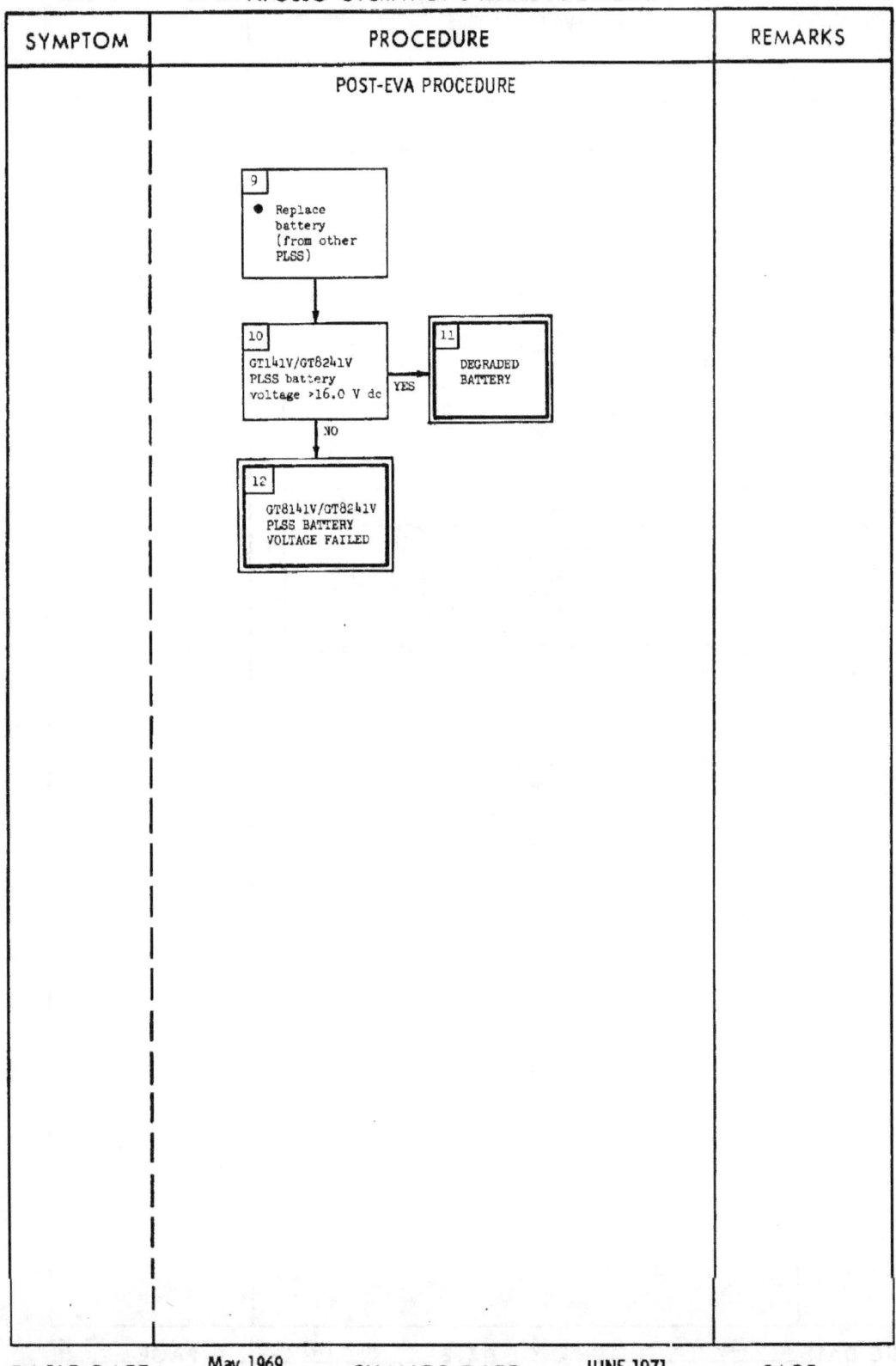	

BASIC DATE May 1969   CHANGE DATE JUNE 1971   PAGE

CSD-A-789-(2) III
APOLLO OPERATIONS HANDBOOK-EMU

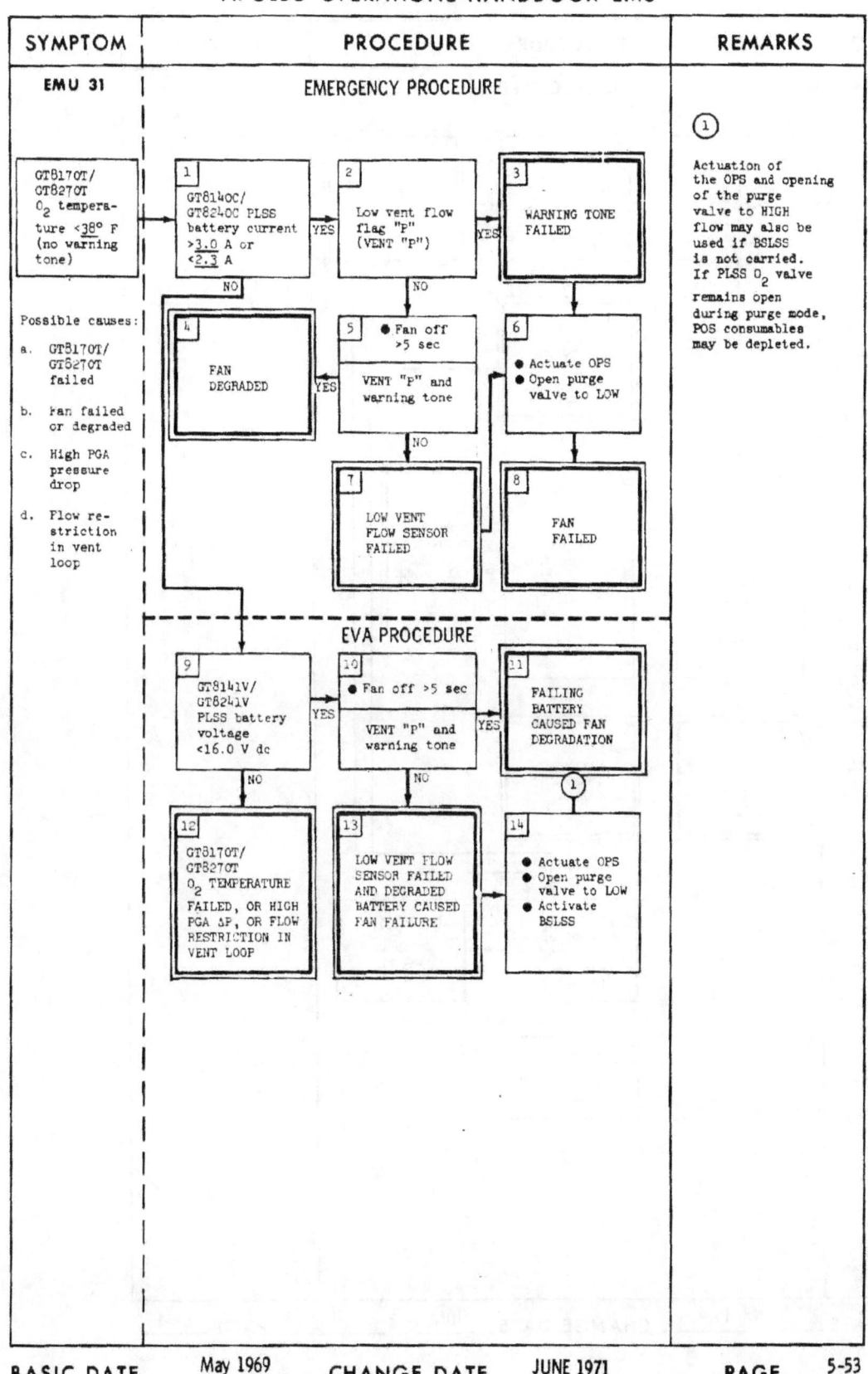

SYMPTOM	PROCEDURE	REMARKS

EMU 31

GT8170T/GT8270T O_2 temperature <38° F (no warning tone)

Possible causes:
a. GT8170T/GT8270T failed
b. Fan failed or degraded
c. High PGA pressure drop
d. Flow restriction in vent loop

Remarks (1): Actuation of the OPS and opening of the purge valve to HIGH flow may also be used if BSLSS is not carried. If PLSS O_2 valve remains open during purge mode, POS consumables may be depleted.

BASIC DATE May 1969   CHANGE DATE JUNE 1971   PAGE 5-53

CSD-A-789-(2) III
APOLLO OPERATIONS HANDBOOK - EMU

SYMPTOM	PROCEDURE	REMARKS
EMU 32	EVA PROCEDURE	

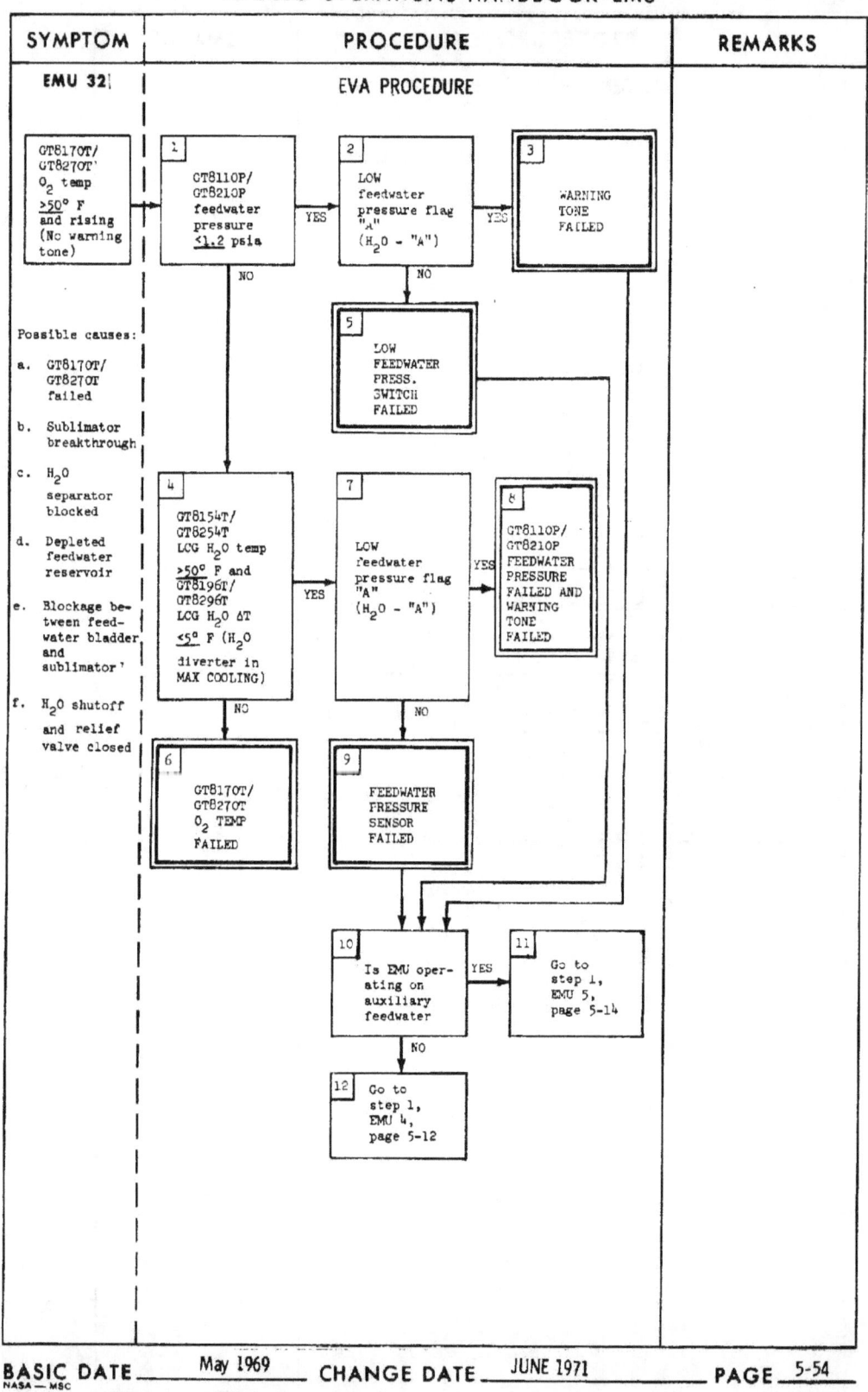

BASIC DATE May 1969   CHANGE DATE JUNE 1971   PAGE 5-54

PROJECT MERCURY

FAMILIARIZATION MANUAL

Manned Satellite Capsule

Periscope Film LLC

NASA
PROJECT GEMINI

FAMILIARIZATION MANUAL
Manned Satellite Capsule

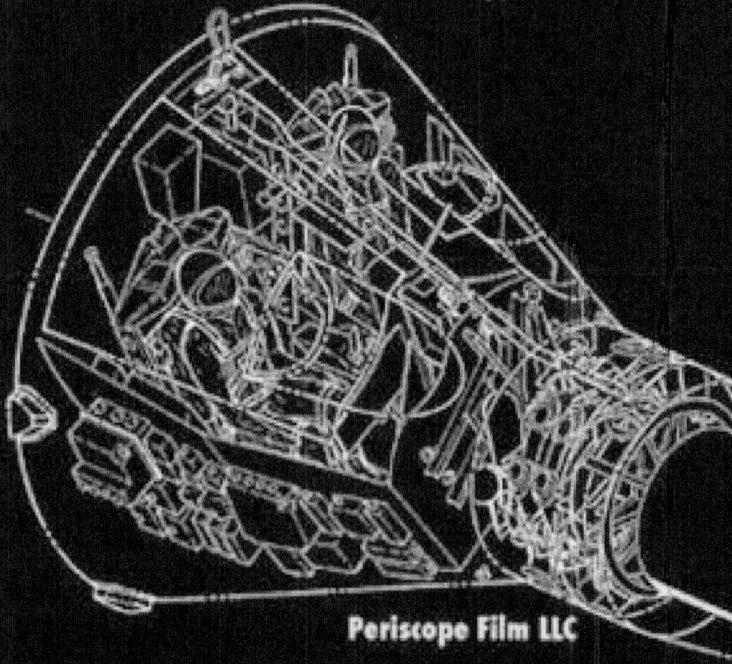

Periscope Film LLC

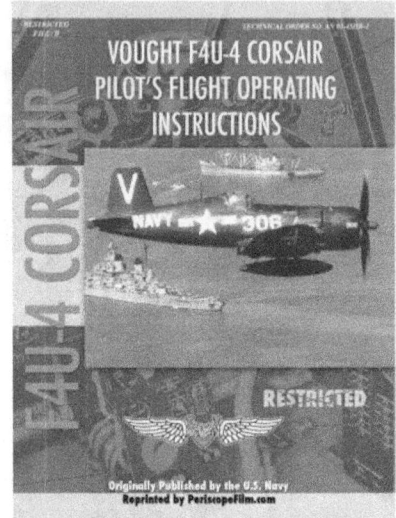
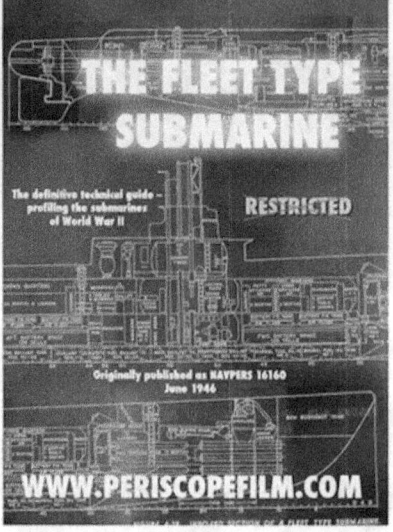

ALSO NOW AVAILABLE FROM PERISCOPEFILM.COM

MMS SUBCOURSE NUMBER 151 — EDITION CODE 3

NIKE MISSILE
and Test Equipment

NIKE HERCULES

DECLASSIFIED

by U.S. Army Missile and Munitions Center and School
Periscope Film LLC

©2012 Periscope Film LLC
All Rights Reserved
ISBN # 978-1-937684-86-0
www.PeriscopeFilm.com

www.ingramcontent.com/pod-product-compliance
Lightning Source LLC
Chambersburg PA
CBHW082112230426
43671CB00015B/2678